Experiments in College Physics II: Physics 1440

COPYRIGHT © 2013-2018 by U.N.T. Physics Department
ISBN 978-1-941563-83-0

All rights reserved. This book or parts thereof may not be reproduced or used in any form or by any means—graphic, electronic, or mechanical, including photocopying, recording, taping, or information storage and retrieval systems—without written permission of the publisher. Making copies of this book, or any portion, is a violation of United States copyright laws.

Fifth Edition
Revised July 2016
Printed in the United States of America

Permission granted by PASCO Capstone to reproduce material in this manual from the *Comprehensive 850 Physics System Experiment Manual* (UI-5813).

For information, contact

University of North Texas
Eagle Images Digital Print Centers
1155 Union Circle #309615
Denton, Texas 76203
940.565.2083 phone
940.369.5311 fax
eagleimages.unt.edu

Disclaimer and Limits of Liability
This manual is the product of the collaborative efforts of many graduate students, staff and faculty members of the Department of Physics at the University of North Texas. Their hard work and creativity are gratefully acknowledged.

The proceeds from this manual supplement student course fees for provision of equipment and services.

The author and publisher do not warrant or guarantee any of the products or procedures described herein. The reader is expressly warned to adopt all safety precautions that might be indicated by the activities described herein and to avoid all potential hazards. By following the instructions contained herein, the reader willingly assumes all risks in connection with such instructions.

Table of Contents

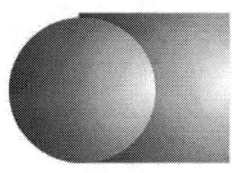

Table of Contents i

Laboratory Information iii

Introduction vii

Understanding and Using Significant Figures xv

Experiment 1: Focal Length of Lens and Mirror 1

Experiment 2: Electrostatic Charging 23

Experiment 3: Electric Field Plotting 45

Experiment 4: Ohm's Law 61

Experiment 5: Series and Parallel Circuit 77

Experiment 6: Capacitance and RC Circuit 95

Experiment 7: Kirchhoff's Circuit Laws 123

Experiment 8: Magnetic Field Mapping 143

Experiment 9: Magnetic Field in a Current Carrying Coil 161

Experiment 10: Induction-Magnet Through a Coil 177

Experiment 11: LRC Circuit-Resonance 191

Experiment 12: Phase Relationships in an LRC Circuit 205

Experiment 13: Reflection, Refraction, Dispersion of Light, and Brewster's Angle 219

Laboratory Information

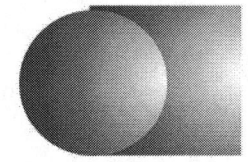

Course Description

PHYS 1440 is the companion laboratory to PHYS 1420; suitable for life sciences majors and pre-professional students. *The laboratory course is separate from the lecture and is separately graded.* The laboratory covers principles and application of electricity, magnetism and light.

Purpose

The purpose of the laboratory course is to give you "Hands On" exposure to the concepts and principles you study in class. An attempt has been made to correlate the labs with the corresponding topics in your lecture; please realize, however, that a perfect match is not always possible. You might get a topic in lab before it has been covered in class.

Attendance

You must attend and complete labs weekly according to the published Weekly Lab Schedule. Please note, it is your responsibility to attend lab as scheduled and to come in and take the corresponding exam. If you miss a particular lab, you receive 0 points for that lab, lab exam and associated lab report (if applicable). **Students who will miss lab due to a school-sponsored event must give notice of the absence to the PIC Director or PIC Assistant at least seven days prior to the event.** It is also your responsibility to be familiar with Physics Instructional Center (PIC) policies on testing and grading. You are responsible for all material in your lab manual including all introductory material.

Grading

Lab grades are based on *accumulated points* acquired by attending and performing lab experiments, taking lab exams and submitting lab reports. Please see the course syllabus for specific guidelines on how your lab grade is calculated.

There are extra labs built into the schedule to accommodate for labs missed due to weather, sickness, family emergency, university closure, etc. Because of this, THERE ARE NO MAKE-UP LABS, LAB EXAMS, LAB REPORTS or EXTRA CREDIT AVAILABLE, except as permitted due to special circumstances as defined by and at the discretion of the PIC Director. In such a case, students <u>must strictly adhere to revised dates and times designated</u> by the PIC Director. You can monitor your points accumulated for the labs during the semester by logging on to UNT's Blackboard Learn system (https://learn.unt.edu) and opening the "My Grades" link on the "Course Content" page. Any grade discrepancies related to lab experiments or exams must be brought to the PIC Director or PIC Assistant <u>immediately within the same week</u> as the lab in question.

Lab Exams

Lab exams can only be taken after completing the lab. **<u>Please note that exams are due within 7 days of completing the related lab experiment</u>**.

Lab Reports

Mandatory lab reports are required and <u>must be submitted through Turnitin in your blackboard</u> within the stipulated timeframe. Please refer to the course syllabus for information regarding specific experiments that require a lab report and the associated due dates.

When submitting your report through Turnitin, you must follow the full submission process. This includes, **uploading** the report, **reviewing** the text on the Preview Panel, and **confirming** and then **submitting** your report. You can be assured that your report has been properly submitted when you see the digital receipt displayed on the screen. The similarity report is also created after successful submission. The digital receipt has a **Submission ID** number which is confirmation that Turnitin has received your paper. If you do not see a digital receipt with a **Submission ID** number, then your paper was not successfully received by Turnitin. A confirmation email is also sent to your UNT email address. Save this email for your records. Any problems encountered during submission must be reported immediately, no later than the next day the University is open for classes.

The following resources are available to assist you in preparing your lab reports:

- "PIC Full Lab Report" guidelines included in the Laboratory Information Section of this lab manual and in your course syllabus.
- "Example Lab Reports" posted to Blackboard
- The "Grading" section in the course syllabus
- If you have any problems or questions, please see PIC Director or PIC Assistant of the Physics Instructional Center, **<u>immediately.</u>**

Lab Procedure Summary

The following is a brief summary of the procedure to follow to get credit for completing your lab experiment and lab exam:
1. Complete the pre-lab questions in your lab manual before coming to lab.
2. Bring your completed pre-lab, student ID, closed-toe shoes and a calculator with your name on it to the lab at the scheduled time. You will turn in your completed pre-lab to a TA for grading as you enter the lab room.
3. Complete your experiment with guidance from the TA and before leaving the lab have the TA verify that your work is complete and correct.
4. <u>As you exit the lab</u>, get the TA to sign your lab book.
5. Take the practice test for the lab exam in the PIC Tutorial Room (PHYS 209).
6. Take the graded lab exam in the PIC Testing Center (PHYS 209AA) using your student ID.

There is a more detailed lab procedure in your syllabus. **You are required to read and familiarize yourself with all the information in the course syllabus and in the Laboratory Information section of this manual.**

Additional Information
- You must purchase a new lab manual. Please check the front cover to assure that it displays the date for the current school year. **Used or outdated books are not allowed**.
- You must have a Photo ID in order to be admitted to the lab class or the lab exam testing room.
- You must wear closed-toe shoes while in the lab. For safety reasons, sandals or flip flops are not allowed.
- Write your name in permanent ink on the front of your lab manual. You can't sell it back to the bookstore at the end of the semester, so go ahead and write on it. It is also wise to put your name on your calculator, as many students accidentally leave theirs behind. Lost calculators without a name on them cannot be returned.

Lab Safety

- Closed-toe shoes or boots are required for all PIC laboratories. No sandals or open-toed shoes will be allowed in the laboratories. All Teaching Assistants, Laboratory Assistants and other PIC personnel are instructed to not admit any student into the laboratories that do not have proper shoes.
- Always tie back long hair.
- No eating or drinking will be permitted while in the lab.
- In some cases you will be instructed by the Teaching or Laboratory Assistant to wear safety goggles. These will be made available to you in the relevant laboratory.
- Some experiments require the use of gas flames. Be sure to tie long hair back and do not turn on the gas higher than is necessary to reach your objectives. If you smell gas, you have the burner turned on too high. Be careful when handling hot objects.

PIC Hours

The PIC Testing Center and PIC Tutorial Lab are non-scheduled, meaning you can come in at almost any time during open hours provided there is space. The PIC Testing Center and PIC Tutorial Lab hours are subject to change based on UNT holidays. Refer to notices posted in PHYS 209 and PHYS 209AA for any changes to the regular hours. The current PIC Testing Center and PIC Tutorial Lab hours are posted to the UNT Physics website, *http://www.phys.edu/PIC*. This information is also provided in your course syllabus.

Tutorial Lab

The PIC Tutorial Lab (PHYS 209) is available as a free service to students taking undergraduate classes in the Physics department. Tutors working in the Tutorial Lab can help you with lab material, lecture homework, and exam preparation.

Testing Center

Lab exams are administered in the PIC Testing Center (PHYS 209AA). Please see the course syllabus for Testing Center hours and exam procedures. Please note that if a blackboard issue occurs during testing, notification must be made **immediately** to the PIC Director/Assistant.

Academic Dishonesty

You must maintain a high ethical standard. If you are caught cheating, you will receive a zero for the lab in question and possibly an F for the semester. All experiment and post-lab questions should be done in the lab room. Any data or questions filled out before you arrive at the lab room, with the exception of the pre-lab questions, will be viewed as cheating and you will receive a zero for that lab and possibly an F for the semester. Any instances of cheating, including but not limited to copying someone else's pre-lab or lab work or lending your lab manual to a friend for them to copy or submitting one's own work done in past semesters, will be reported to the Office of Student's Rights and Responsibilities.

Please refer to the UNT Student Academic Integrity policy:
https://policy.unt.edu/sites/default/files/06.003.pdf.

Introduction

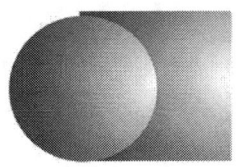

Science is the organization of apparently unrelated facts into a useful whole. Science goes beyond verifying what we understand. We use science to challenge what we "know" and develop methods to prove or disprove our hypothesis.

Physics is that branch of science that serves as the foundation for all other scientific and engineering disciplines. Laboratories for physics are the classrooms where the relationships of physics and science in general, are illustrated in a manner that encourages and even depends upon your participation. It is the goal of the physics laboratories to introduce you to the relationships of physics. Lab exercises will provide the hands-on experience necessary to appreciate the methods of science and scientific discovery. In order to understand physics, and science in general, you must be involved in the laboratory.

The physics laboratory is designed to do more than simply illustrate proven relationships. Its ultimate goal is to provide you with the tools and experience necessary for you to pursue your own discovery. In these labs you will explore some of the fundamental concepts of physics. More importantly, you will be introduced to some of the tools of science and experience the methods of scientific discovery.

The Lab

Role of the Student

For many students the concepts and relationships discussed in lecture are difficult to visualize and have no application outside of the physics course. Exercises in this lab manual are designed to help you bridge the gap between concept and application. The labs have been carefully chosen and structured so that the concepts discussed in lecture can be demonstrated as well as verified. It is your responsibility to accurately complete the exercises to successfully illustrate the concepts of physics.

Lab Exercises

This manual contains lab exercises based on class discussion. Each exercise requires a certain level of familiarity with the physical concepts to be studied. Study the exercises before coming to the lab and review any additional material as may be necessary. Each exercise is designed and written to minimize the amount of outside reference material required.

However, a careful reading of the material covered in lecture will complement the laboratory experience. Each exercise contains the following:
- Overview of the physical concept to be explored
- What you should achieve and learn having successfully completed the lab
- A list of required equipment
- A step by step procedure complete with all necessary data tables, illustrations, and room for calculations
- Analysis designed to reinforce and review what you learned

Procedure

Collecting Your Data

Laboratory experiments are ultimately about collecting data, organizing it in a concise manner, and inferring some result. The data will lend support for or against a given hypothesis. It is necessary to have a consistent and uniform procedure for reading instruments. Tabulating and mathematically manipulating data is necessary for the experimental results to be easily understood. The way that we report and use the data gathered in these labs needs to illustrate the magnitude of some observed quantity and communicate how confident we are about or how much significance can be placed on our results. This process begins with how we use our instruments. What follows is how we determine the level of precision and accuracy that we expect from our instruments and measurements.

The use of instruments and the reporting of data in a laboratory requires a careful definition of four terms; ***accuracy, precision, resolution,*** and ***sensitivity***.

Accuracy of a measurement is its agreement to some known or true value. Accuracy is often a measure of whether or not an experimenter is using the appropriate equipment properly set up. Using a meter stick to measure the thickness of a sheet of paper will give inaccurate results. Using an instrument not properly calibrated or set will also give poor accuracy.

Precision is a measure of how reproducible the value is. The precision of a measurement is its agreement to previous measurements made by the same person using the same instrument. Typically, equipment properly used in a consistent and careful manner will have high precision within the resolution of the instrument.

Resolution is the limit or smallest value that can be reasonably read on the instruments scale. Resolution is typically limited by the smallest markings or gradation of the instrument. A meter stick will typically have a resolution of about .5 millimeters. A micrometer however will typically have a resolution of about .01 to .005 of millimeters.

Sensitivity is the limit or range of values in which instruments can reasonably be expected to give accurate results. Measuring the thickness of a piece of paper with a meter stick will always give poor results. Measuring the time it takes to blink an eye using a wrist watch is extremely difficult since the event takes place much quicker than the sensitivity of the watch.

Recording Your Data

The magnitude or greatness of a value, properly recorded to indicate the appropriate significance of the measurement, is not yet a true physical quantity. In order for your data to have true significance it must also express what the quantity is physically. This is done by assigning each value measured an appropriate unit. In these labs you will be working within the metric system, or System International (SI). In this system the three basic units of length, mass, and time are the meter, kilogram, and second, respectively. Additionally we will be using the ampere for current, volt for potential difference, and the Kelvin or degree Celsius for temperature.

The *mks* (meter, kilogram, second) convention is simply a matter of convenience when measuring and reporting physical quantities. An alternate conventions used in some labs are the *cgs* (centimeter, gram, second) units. The decision to use either the *mks* or *cgs* convention is actually determined by the relative magnitude of the quantities measured. At times it is simpler to work with centimeters and grams than it is to use the meter and kilogram. However, once a decision is made to use either the *cgs* or *mks* convention, it is crucial that you remain within that convention. The data tables for each exercise are annotated with the recommended unit of that particular entry. Any additional discussion of the choice of units for a particular lab will be detailed in the procedure section of that lab exercise.

At times it may be necessary to record a measurement using kilograms as your choice of units but the value itself is on the order of a hundredth of a gram. You write in your data table .00017 kilograms. Although the representation of this number is entirely correct, it does get rather cumbersome carrying around all of those zeros. The solution to this is to represent the value in scientific, or powers-of-ten, notation. To express your value in this manner, insert a decimal point after the first non-zero figure then multiply by a power of ten to locate the true decimal point. In locating the true decimal point, count the number of places that the decimal point was moved in order to represent it in scientific notation. In our example, we count four places to the right. Our value of .00017 kilograms would then be expressed as 1.7×10^{-4} kilograms. A value of 17000 kilograms would become 1.7×10^4 kilograms. If you move your decimal point to the right your exponent is a negative number, to the left the exponent is a positive number. Another advantage to scientific notation is the ease in which significant figures are determined for your data. When using scientific notation, you only write those values that are significant. Our value of .00017 kilograms only has two significant figures, therefore we express it as 1.7×10^{-4} kilograms. A more detailed discussion on significant figures will be given in a subsequent section.

Using the appropriate units and recording your data, is simply a matter of noting the value you read from the instrument. Read the instrument to the *resolution* of the tool. Keep in mind that your instrument needs to have sufficient *sensitivity* to give meaningful results. You need to be careful and consistent to maximize the *precision* of your measurements. You must use the proper tool, calibrated to be as *accurate* as possible. It's that simple. If you apply these concepts every time you take a measurement and write a value with its appropriate units in the data tables, every lab exercise can be successfully completed.

Data Analysis

After obtaining all data, you are ready to start evaluating and determining whether or not it agrees with the expected results. You will also be trying to determine how well your data are in agreement with what you expected. Determining how well your data are in agreement to some known standard and assigning a mathematical value to that agreement is called quantitative analysis. However you need to do a qualitative analysis first.

Qualitative Analysis is asking yourself if the data you just measured or the value you just calculated makes sense. For example, try asking yourself "is this about what I expected to get?" If you are measuring the velocity of a small child riding a bike and you get a value of 29 meters per second (65 mph), something is probably wrong. If, however, that child is in a car traveling down a highway, 29 meters per second may be reasonable. If you are not sure that the measured or calculated values are reasonable, ask your instructor. Qualitative analysis is a good way to catch mistakes in the experimental setup and data collection to avoid repeating the entire experiment.

Quantitative Analysis is where you statistically determine how accurate and precise the data are. The extent to which you analyze the experimental results will normally be stated in the procedure section of your lab. This may be as simple as finding a percent difference from some known or standard value. It could involve comparing your experimental error to the error you might reasonably expect from your setup and available equipment. Regardless of the extent to which you analyze your data, it all begins with proper use of significant figures.

In scientific work, most numbers are measured quantities and are not exact. All measured quantities are limited in significant figures (SF) by the resolution of the instrument used to make the measurement. The measurement must be recorded to show the resolution of the instrument—no more, no less. Calculations based on the measured quantities can have no more (or no less) precision than the measurements themselves. The results of your calculations must be recorded to the proper number of significant figures. To do otherwise is misleading and improper.

Errors

It is important to be aware of some of the sources of experimental error and some basic statistical methods of determining the value or values that best represent the system in question. Sources of error that affect the outcome and results of your lab normally result from one of two general categories: random error that you have little control over but which can be addressed using statistics, and systematic error that you either can control, minimize or measure.

Random Error is the error that occurs at the limit of the resolution of your instruments. It is random error that necessitates the proper use of significant digits. Every instrument has a limit as to how well it can record a value. When listing a value to three significant digits in data tables, you are making a statement that implies you are unsure of what the next figure should be. Is my last reported value close to being rounded up or down? On average you are just as likely to be at one end of the scale as the other. The way we address this issue is by taking an average of several values. The Standard Deviation of your values will determine the "spread" or variance of your data.

Systematic Error is the error we create. Systematic error includes such things as consistently reading a scale too high or too low, having an instrument not calibrated or set properly, trying to use an instrument outside of its range of sensitivity, improper experimental setup, or poor assumptions as to whether the setup will illustrate the concept that you are trying to explore. Systematic error can be minimized by the proper use of equipment. Each time you use a new piece of equipment, make sure it is properly calibrated. The best and easiest way to calibrate an instrument is to make sure it reads "zero" when it should. If zeroing the instrument is not appropriate, you might try checking the instrument against some known standard. To minimize a personal bias to read an instrument too high or too low, let different people read the measurement. Vary your method. When measuring the length of an object, do so at several different locations. While systematic error may still be present, you can minimize it so that it is smaller than your random error.

The distinction needs to be made at this point between "error" and performing the experiment improperly. When we talk about error, we are referring to the errors that affect our data at the limit of our ability to use the equipment or at the limit of the resolution of the equipment. You might be tempted to say my error was that, "I did something wrong." Some call this "personal" error. Granted, personal error will certainly affect your final results. However, it should be corrected as soon as it is discovered and **that portion of the experiment repeated.**

You also need to know how to compare an experimental value to either another experimental value or to some known or accepted value. When comparing an experimental result to another experimental result, calculate a percent difference. The percent difference is the absolute value of the difference between the values that you are comparing divided by the average of the two values,

$$\text{percent difference} = \frac{|\text{value1} - \text{value2}|}{\text{average value}} \times 100\%$$

When comparing your experimental results to some known or accepted value, we will calculate a percent error. The percent error is the absolute value of the difference between the two values divided by the accepted value,

$$\text{percent error} = \frac{|\text{exp value} - \text{accepted value}|}{\text{accepted value}} \times 100\%$$

The difference in these two methods is that with the percent error we know or at least have a much better understanding of the accuracy of the accepted value. We know the value well enough so we can say it is essentially exact in comparison to our experimental value. Therefore, we can calculate an absolute deviation. When calculating a percent difference, we have no reason to believe one value is any better or worse than the other. Therefore, we will find the deviation from the average of the two values which should better represent the actual value.

Plotting of Data

In physics one of the best methods for visualizing and interpreting data is with the use of graphs. Graphs allow you to transfer data from a table onto a scaled, clearly labeled grid. The primary purpose of graphing data is to understand better the functional relationship between variables. Once your data is properly graphed, it is simple to predict an approximate value for one variable given you know the other, or to verify extreme relationships between your variables.

Most physical relationships in nature can be graphically analyzed using one of three different types of plotted curves. Fortunately, all three of these can be graphed as a linear relationship with only minor mathematical manipulation. They are:

Linear Relationships —such as $y = mx + b$
Power Relationships —such as $y = ax^m$
Exponential Relationships —such as $y = be^{mx}$

Before you start graphing your data, it is important to properly scale and label your graph paper. Plot the independent variable along the x-axis and the dependent variable along the y-axis. For example, our relationship $y = mx + b$ has the form such that y is our dependent variable and x is our independent variable. Is the distinction between independent and dependent variables just a matter of how our equation is expressed? Could we have rearranged the equation where x was dependent on y? Yes, but the distinction between dependent and independent variables is actually a consequence of what you can actually measure in the laboratory. For the relationship $y = mx + b$, we were evidently able to vary x and measure y. Therefore, y is dependent on x. Once you have decided which variable belongs to which axis, you must choose a scale.

There are two important steps in scaling your graph, first, use a scale that allows your plotted curve to fill as much of the graph paper as possible. Secondly, use a scale that is convenient and easy to read such as one that increases as a factor of either 1, 2, or 5 of your units of measure.

Example: Your data ranges from 1.103 meters to 2.047 meters. The graph paper is divided into twenty, evenly spaced grids. If you subtract your two extreme data points, your values encompass a range of .944 meters. This value evenly distributed among all twenty divisions of your graph paper would give .0472 meters per grid, a difficult number with which to work. Make the graphing simpler by rounding up the .0472 to the next multiple of 1, 2, or 5. In this case you would choose .05. Label your graph axis starting with 1.10 meters and continue in increments of .05. Your next value would be 1.15, 1.20, and so on. Once you have plotted all your points, be sure to label all axes with the appropriate units, give your graph an appropriate title, and include any additional scaling information as required.

In the linear relationship y = mx + b,

- y and x are variables plotted on your graph
- m is a constant value that represents the slope of your graph
- b is the y value when x = 0, or the y intercept.

The slope of your graph is the ratio of the change in the y values to the change in the x values. The slope of your graph is determined from the "line of best fit" that you draw on your graph paper. The line of best fit is a straight line you think best depicts the average of your values.

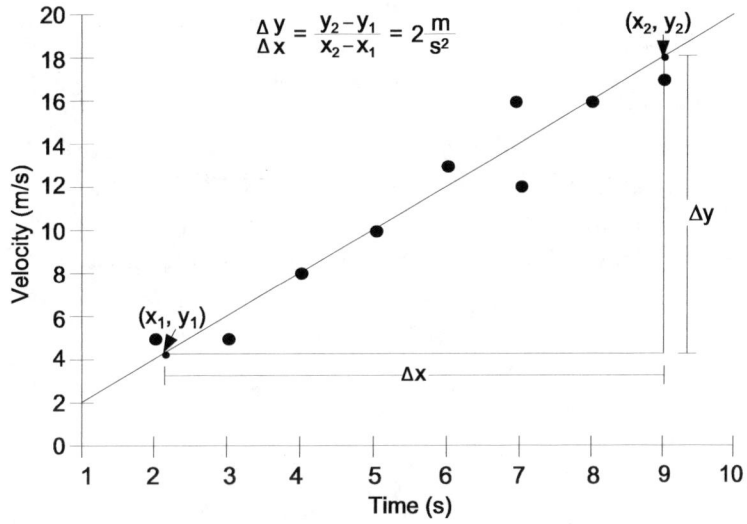

Figure 1. "Line of Best Fit"

The power relationship is difficult to graph and difficult to work with in the form

$$y = ax^m$$

However, if you take the Log of both sides of the equation you obtain;

$$\text{Log}(y) = \text{Log}(ax^m)$$

rearranging gives

$$\text{Log}(y) = m\,\text{Log}(x) + \text{Log}(a).$$

In this form, if you plot Log(y) versus Log(x), the slope of your graph will be m. This type of plot is commonly referred to as a Log-Log plot. If m were always an integer it would be almost as easy to plot y versus x^m. The slope would then become the value of a. This method is certainly satisfactory as long as you are sure that m should be an integer.

The third of our relationships, the exponential relationships, can be handled by taking the natural log of the equation $y = be^{mx}$;

$$\ln(y) = \ln(be^{mx})$$

rearranging gives

$$\ln(y) = \ln(e^{mx}) + \ln(b)$$
$$\ln(y) = mx + \ln(b)$$

In this form, a plot of ln(y) versus x would give a linear graph with slope m and y intercept of ln(b). A graph of this type is a semi-log graph.

Becoming proficient at graphing your data and properly interpreting the results is a matter of practice. The only way to truly understand and appreciate the usefulness of graphical analysis is to work your way through some labs. The most important point of any graphing exercise is to be neat. When graphing, clearly label your axis and name your graph. When labeling your axis, use the appropriate units as well as magnitudes. The "line of best fit" is really just an approximation of what you think the average values represent. The slope that is determined from your "line of best fit" does have physical significance and must be calculated with the units.

Understanding and Using Significant Figures

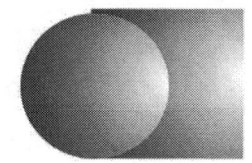

In scientific work, most numbers are measured quantities and thus are not exact. All measured quantities are limited in significant figures (SF) by the precision of the instrument used to make the measurement. The measurement must be recorded in such a way as to show the degree of precision to which it was made - no more, no less. Calculations based on the measured quantities can have no more (or no less) precision than the measurements themselves. The answers to the calculations must be recorded to the proper number of significant figures. To do otherwise is misleading and improper.

Determining Which Figures are Significant

- Non-zero integers are always significant.
    example: 23.4g and 234g both have 3 SF

- Captive zeroes, those bounded on both sides by non-zero integers, are always significant.
    example: 20.05 has 4 SF; 407 has 3 SF

- Leading zeros, those not bounded on the left by non-zero integers are never significant. Such zeros just set the decimal point; they always disappear if the number is converted to powers-of-10 notation.
    example: 0.04g has 1 SF; 0.00035 has 2 SF. They can be written as 4×10^{-3} and 3.5×10^{-4} respectively.

- Trailing zeros, those bounded only on the left by non-zero integers may or may not be significant.
    example: 45.0L has 3 SF; 450L has only 2 SF; 450.L has 3 SF.

    Note: To clarify whether a trailing zero is significant, it is preferable to use scientific notation to express the final answer.
    example: 450.L can be expressed as 4.50×10^2 or 4.50E02 whereas 450 L would be expressed as 4.5×10^2.

- Exact numbers are those not obtained by measurement but by definition or by counting numbers of objects. They are assumed to have an unlimited number of significant figures.

Multiplication and Division Involving Significant Figures

Calculations involving only multiplication and/or division of measured quantities shall have the same number of significant figures as the fewest possessed by any measured quantity in the calculation.

    example: 14.0 x 3 = 40, not 42, because one of the multipliers has only one SF.
    example: 14.0 x 3.0 = 42, because one of the multipliers has only two SF.
    example: 14.0 / 3 = 5, not 4.6, because the denominator has only one SF.

Addition and Subtraction Involving Significant Figures

Calculations where measured quantities are added or subtracted shall correspond to the position of the last significant figure in any of the measured quantities. That is, the final answer is only as precise as the decimal position of the least precise value. The number of significant figures can change during these calculations.

    example: 14.16　　　　　　example:　　46.6
               + 3.2　　　　　　　　　　　　　　+ 5.72
               17.36　　　　　　　　　　　　　　52.32
               17.4 is the correct answer　　52.3 is the correct answer

Combined Calculations

In calculations involving addition/subtraction and multiplication/division, significant figure guidelines must be applied *after each calculation involving addition or subtraction*.

    example: (3.2 x 4 x 0.035 / 7) + (12 x 0.5) =
               0.06 + 6 = 6

Tips for Rounding Off Numbers

A number is rounded off to the desired number of significant figures by dropping one or more digits to the right. The following guidelines should be observed when rounding off numbers.

- When the first digit dropped is less than 5, the last digit remains unchanged.
- When the first digit dropped is more than or equal to five, the last digit retained is increased by 1.
    examples:　　243 → 240　　　　17.9 → 18

Conclusion

The precision of the instruments used in collecting data determines the degree to which your results are accurate. Significant figures provide an easy way of indicating that accuracy. Using these guidelines assures that the data resulted from your procedures is not only reproducible, but also allows an observer to understand the degree to which your data is accurate.

Last/First Name (print): _____

PHYS 1440-Section _____   Username ID (xxx9999): _____

Experiment 1:
Focal Length of a Lens and a Mirror

Note: Be certain to read the whole experiment to answer these questions.

Pre-lab, Part A

1. During the first part of this lab, we want to determine how the object distance is related to what two quantities?

2. How will a convex lens change the image of an object? Explain your answer.

Pre-lab, Part B

3. Name two examples of spherical mirrors that we will use in our experiment.

4. Why are images that are formed by a convex mirror called virtual images?

5. Define the following, and give the letter which we will abbreviate them by:

   Center of curvature:

   Vertex:

   Focal Point:

   Radius of curvature:

   Focal length:

6. If the radius of curvature is known to be .08m for a particular concave mirror, what is the focal length of this mirror?

Last/First Name (print): _____

Experiment 1:
Focal Length of a Lens and a Mirror

Introduction A: Focal Length of a Lens

The purpose of this activity is to determine the relationship between object distance and image distance for a thin convex lens. Use a light source, optics bench, lens, and viewing screen to measure object distance, image distance, and image size.

Theory A: Focal Length of a Lens

The behavior of light through a thin lens is utilized every day in simple devices that we use. Some of these devices include magnifying glasses, telescopes, and binoculars. Even the human eye takes advantage of the lens found in its outer surface, focusing images so we can see clearly.

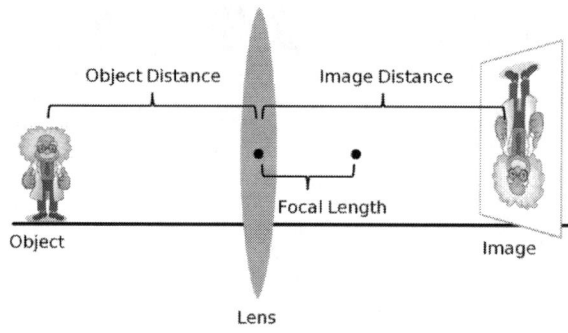

Figure A1: Image Formation from a Lens

On a small scale, the behavior of light as it passes through a thin lens can be described using Snell's Law, the angle at which light is incident on the lens surface, and the index of refraction of the material the lens is made out of. However, using a more macroscopic scale, we can characterize this behavior using spatial measurements and simple geometry that involves the distance between the object of interest and the lens, known as "object distance", the distance between the lens and the object image, known as "image distance", and a special characteristic of the lens known as "focal length".

The goal of this lab will be to explore the behavior of light and the images it forms as it travels through a thin lens, and collect data that will help determine how image distance is related to both the object distance and focal length of a thin convex lens.

$$\frac{1}{f} = \frac{1}{d_o} + \frac{1}{d_i}$$

Equipment:

1	Basic Optics Light Source	OS-8470
1	Optics Track	OS-8508
1	Basic Optics Viewing Screen	OS-8460
1	Convex Lens Set	OS-8456
1	Concave/Convex Mirror	OS-8457
1	Concave Mirror Accessory	OS-8532
1	Ruler	
1	PASCO Capstone	UI-5400

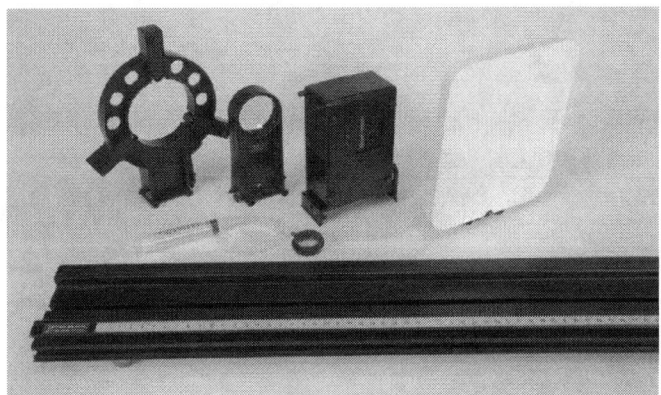

Figure A2: Equipment

Setup A: Convex Lens with Fixed Focal Length

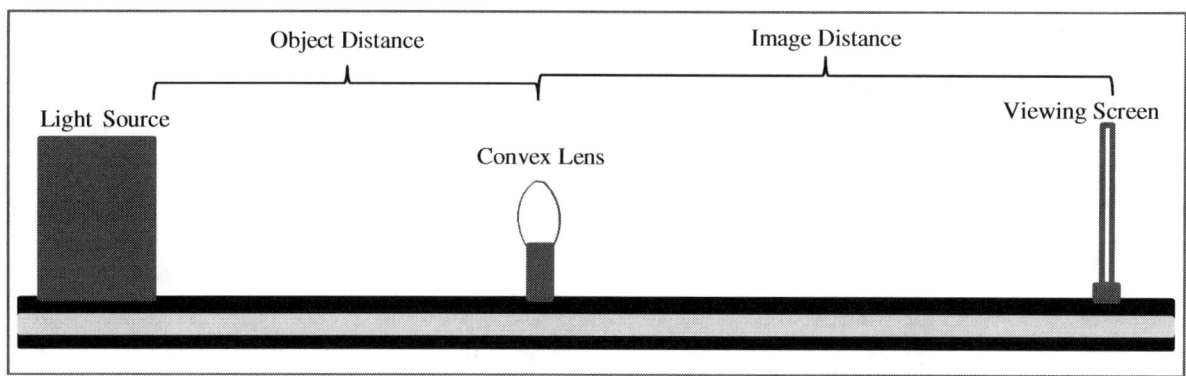

Figure A3: Convex Lens with Fixed Focal Length Setup

1. Make certain the light source is at the zero point on the optics track with the crossed-arrow pattern facing down the length of the track.
2. Mount the viewing screen at the other end of the bench with the flat surface facing the light source.
3. Mount the 100-mm glass convex lens to the optics track 500 mm from the light source. Use the metric scale on the optics track and the marks at the base of the light source and lens mount (as in Figure A4) to help set this distance.

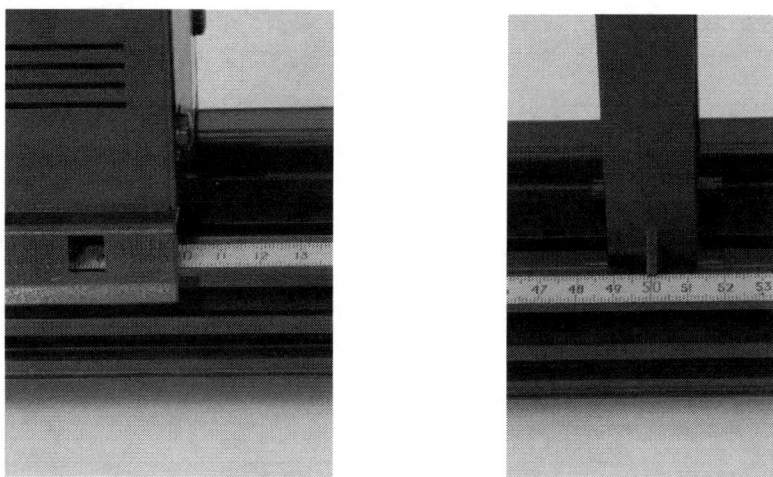

Figure A4: Track Marker

Last/First Name (print): _____

PHYS 1440-Section _____ Username ID (xxx9999): _____

Procedure A: Convex Lens with Fixed Focal Length

1. The image of the crossed-arrow target should initially appear out of focus. Adjust the position of the viewing screen forward or backward until the image of the crossed-arrow target is in focus.
2. When the image is in focus, measure the distance between the lens and the viewing screen. Record this image distance in Table A1.
3. Use the ruler to measure the diamerter of the image. Record this as Image Height in Table A1. Use negative values to indicate that the image is inverted.
4. Adjust the position of the 100-mm convex lens to the next object distance listed in Table A1 of 475 mm.
5. Adjust the position of the viewing screen until the image of the crossed-arrow target is in focus, and then record the image distance and height in Table A1.
6. Repeat the same procedure for each object distance value in Table A1. Record all values in Table A1.

Table A1: Object and Image Distances

Object Distance d_o (mm)	Image Distance d_i (mm)	Image Height h_i (mm)
500		
475		
450		
425		
400		
350		
300		
250		
200		

Last/First Name (print): _____

PHYS 1440-Section _____ Username ID (xxx9999): _____

Analysis A1: Relationship between Image and Object Distance to Focal Length of a Lens

The graph that follows shows image distance (mm) as a function of object distance (mm).

1. What value does the image distance approach as the object distance becomes larger (approaches infinity), and what is the significance in this value?

The graph of image distance versus object distance isn't linear; however, by clicking on either axis label you can select QuickCalcs that will quickly manipulate either axis measurement by operating on it mathematically. Use a QuickCalc of $\frac{1}{y}$, and $\frac{1}{x}$.

2. Apply a linear fit to your data and write the y = mx + b equation that describes the line below (round m to the nearest whole number). Replace y and x using the variables from your answer to the previous question.

3. What is the significance of the y and x-intercepts in your graph/equation?

4. Based on your data and the answers to the previous questions, what is the equation that relates object distance, image distance, and focal length of a thin lens?

5. Make a rough sketch of the linearized data here, showing the x and y intercepts.

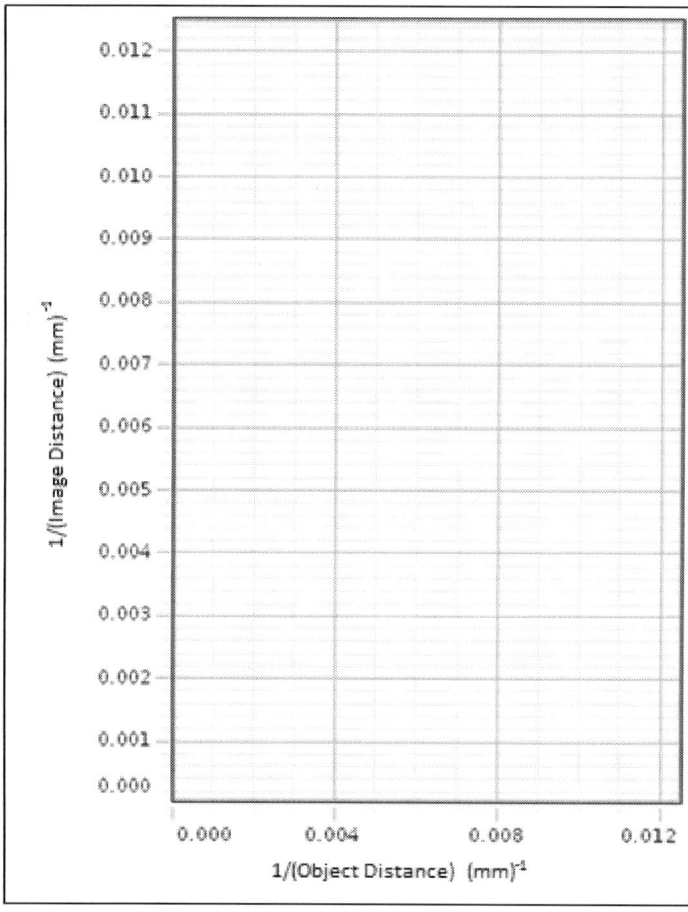

Last/First Name (print): _____

PHYS 1440-Section _____ Username ID (xxx9999): _____

Analysis A2: Magnification

Magnification is a measure of how much an image has been magnified from its original size. Mathematically, magnification is the ratio of image height to object height:

$$Magnification = \frac{-Image\ Distance}{Object\ Distance} = \frac{Image\ Height}{Object\ Height}$$

A negative sign of magnification represents inverted image, while a positive sign represents upright image.

1. Click in the first cell in the Magnification column. This will reveal the calculation for magnification at the top of the table.

2. Use the ruler to measure the height of the object, and then enter that value into the magnification equation at the top of the Table A2. Press enter when the equation is complete; data values will auto-populate the magnification column in the table. Sketch your data in the magnification vs. object distance, and magnification vs. image distance graphs that follow Table A2.

Table A2: Magnification	
Object Distance d_o (mm)	Magnification
500	
475	
450	
425	
400	
350	
300	
250	
200	

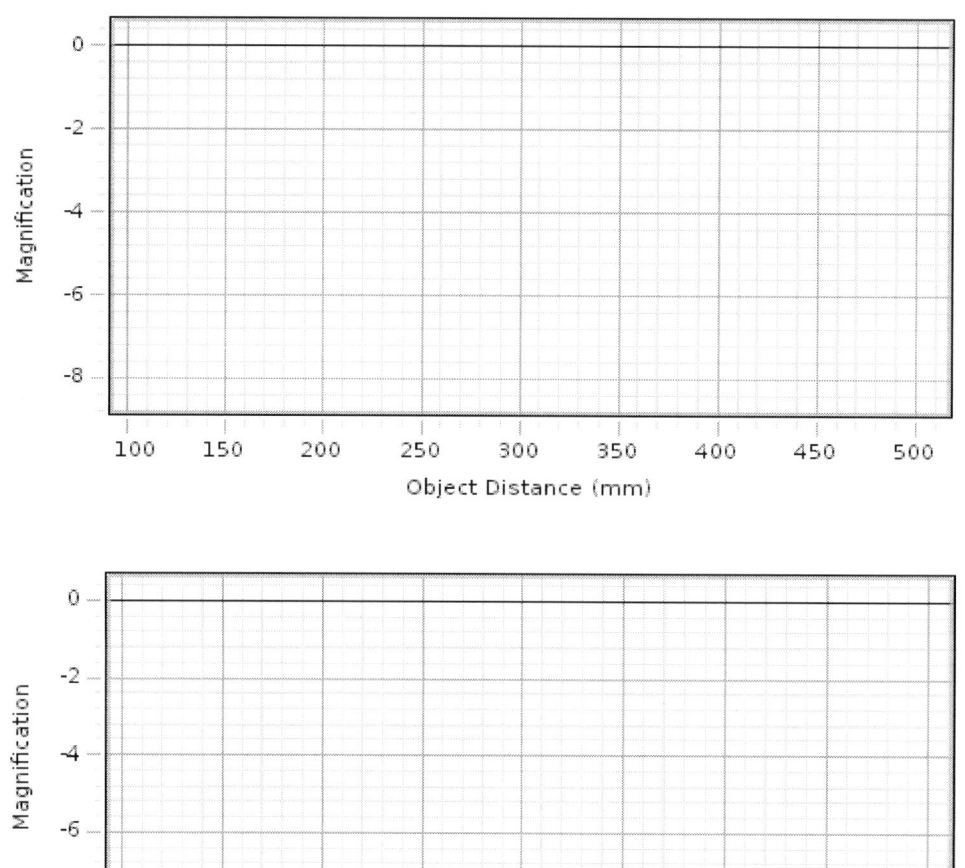

The two graphs above show magnification as a function of object distance (mm), and image distance (mm).

3. How is magnification related to object distance (proportional, inverse, etc.), and at what object distance is the magnification approximately equal to -1 (object height = image height)?

Last/First Name (print): _____

PHYS 1440-Section _____ Username ID (xxx9999): _____

4. How is magnification related to image distance (proportional, inverse, etc.), and at what image distance is the magnification equal to -1? How does this value compare to the object distance at the same magnification?

5. Based on the responses to the previous two questions, what is the mathematical relationship that relates magnification to object and image distance?

6. What value does magnification approach as object distance approaches the focal length of the lens (100 mm)? What do you think happens to the image when the object distance is less than the focal length of the lens?

Introduction B: Focal Length of a Concave Mirror

The purpose of this activity is to measure the focal length of a concave mirror. Use a light source, concave mirror, and half screen accessory on an optics bench to measure the focal length of the concave mirror.

Theory B: Focal Length of a Concave Mirror

Concave and convex mirrors are examples of spherical mirrors. Spherical mirrors can be thought of as a portion of a hollow sphere which was sliced away and then silvered on one of the sides to form a reflecting surface. Concave mirrors are silvered on the inside of the sphere and convex mirrors are silvered on the outside of the sphere.

If a concave mirror is thought of as being a slice of a sphere, then there would be a line passing through the exact center of the sphere. This line is known as the principal axis. The point in the center of the sphere from which the mirror was sliced is known as the center of curvature and is denoted by the letter C in the diagram. The point on the mirror's surface where the principal axis meets the mirror is known as the vertex and is denoted by the letter A in the diagram. The vertex is the geometric center of the mirror. Midway between the vertex and the center of curvature is a point known as the focal point; the focal point is denoted by letter F in the diagram. The distance from the vertex to the center of curvature is known as the radius of curvature (abbreviated by "R"). The radius of curvature is the radius of the sphere from which the mirror was cut. Finally, the distance from the mirror to the focal point is known as the focal length (abbreviated by "f").

The focal point is the point in space at which light incident towards the mirror and travelling parallel to the principal axis will meet after reflection. Figure B2 depicts this principal. In fact, if some light from the Sun was collected by a concave mirror, then it would converge at the focal point. Because the Sun is such a large distance from the Earth, any light rays from the Sun which strike the mirror will essentially be travelling parallel to the principal axis. As such, this light should reflect through the focal point.

The image formed by the light rays as they converge at the focal point of a concave mirror can appear on a card placed at the focal point. Because this image can form on a physical object, the image is known as a real image. This is in contrast to the other type of spherical mirror – a convex mirror – in which the light rays appear to converge at a point behind the mirror. Such images cannot be formed on a physical object, and are known as virtual images.

For any spherical mirror, the focal length f is related to the radius of curvature R of the mirror by:

$$f = \frac{1}{2}R$$

where R is always a positive value for a concave mirror.

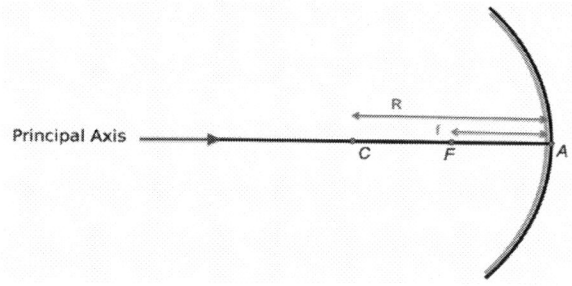

Figure B1: Concave Mirror

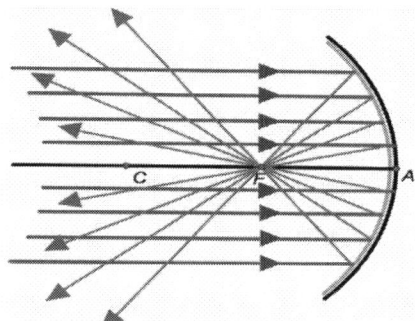

Figure B2: Focal Point of the Concave Mirror

Setup B

1. Mount the Light Source at 0mm end of the Optics Bench. Place the Concave Mirror at the 250mm mark of the Optics Bench.
2. Position the Light Source so the crossed arrow target is aimed at the Concave Mirror and the concave surface of the mirror faces the light source.
3. Place the Half-Screen a few centimeters in front of the Concave Mirror (between the mirror and the Light Source) as shown in figure B3.

Figure B3: Equipment Setup

Last/First Name (print): _____

PHYS 1440-Section _____   Username ID (xxx9999): _____

Procedure B:

1. Move the Half-Screen closer to or further from the Concave Mirror until the reflected image of the crossed arrow target on the white screen is focused.
2. Measure the distance between the position indicators on the Half-Screen and the Concave Mirror. Record this as image distance.
3. Move the Concave Mirror to 500mm mark of the Optics Bench.
4. Repeat steps 1-2 until the image distance values for each object distance on the table is recorded.

Figure B4: Crossed Arrow Target Image on the White Screen

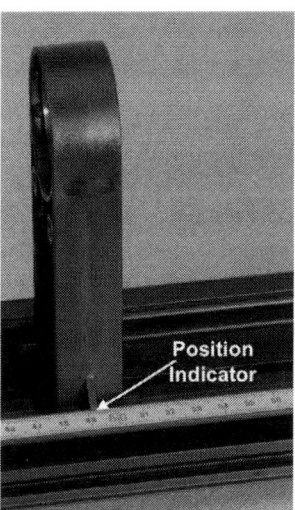

Figure B5: Position Indicator on the Concave Mirror

Object Distance (mm)	Image Distance (mm)
250	
500	
750	
1000	
1150	

Last/First Name (print): _____

PHYS 1440-Section _____ Username ID (xxx9999): _____

Analysis B: Focal Length of a Concave Mirror

1. Record the focal length of the Concave Mirror which is printed below the mirror.

2. Knowing the focal length, which trial gave you the closest measurement to the focal length? What was the percent difference between that trial and the focal length?

3. How might you determine the focal length more accurately?

4. What is the orientation and size of the target compared to the image itself?

1440 Questionnaire: Experiment 1: Focal Length of Lens and Mirror

Do not put your name on this page. Hand in this page separately.

1. TA name(s):

2. What one thing did you like best about this laboratory?

3. What one thing did you like least about this laboratory?

4. What one thing would you change in this laboratory?

5. What one thing would you leave the same?

Additional Comments?

Last/First Name (print): _____

PHYS 1440-Section _____ Username ID (xxx9999): _____

Experiment 2:
Electrostatic Charging

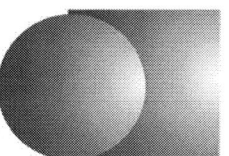

Note: Be certain to read the whole experiment to answer these questions.

Pre-lab, Part A

1. Briefly explain what it means when an object is negatively charged, in terms of proton/electron balance.

2. In charging, what is being transferred between objects? Also, list the three different types of charging an object, as explained in the theory section.

3. What does the law of conservation of charge say?

4. If an object has a concentration of positive charges on one side, and a concentration of negative charges on the other side, but the object is electrically neutral, what can be said about the object?

5. If a neutral conductor comes in contact with a positively charged object, what will the distribution of charge look like? To answer this, roughly finish the diagram below with + and − symbols to represent charge distribution.

Pre-lab, Part B

6. The force between two point charges obeys which law?

Last/First Name (print): _____

Experiment 2: Electrostatic Charging

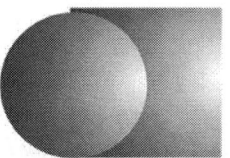

Introduction

The purpose of part A of this experiment is to compare and contrast the results of three different methods of charging: (1) rubbing two objects together; (2) touching a charged object to a neutral one (charging by contact); and (3) grounding a neutral object while it is polarized (charging by induction).

Part A will also demonstrate the law of conservation of charge.

The purpose of part B of this experiment is to investigate how charge distributes on the outer surfaces of a spherical conductor and on the outer surface of a non-spherical conductor. The charge distribution inside the spherical conductor will also be examined.

Theory A

Electric Charges
An electric charge is a fundamental property of nature. It comes in two types, positive and negative. Positive charge is the type of charge carried by protons. Negative charge is the type of charge carried by electrons. As nearly as can be measured (better than 1 part in 10^{30}), the magnitude of the charge on an electron is the same as the magnitude of the charge on a proton. Atoms normally have the same number of protons and electrons and this balance of charges makes them electrically neutral. Most objects are found in this neutral state. For an object to be positively charged, it has to have more protons than electrons. For an object to be negatively charged, it has to have more electrons than protons, disturbing the neutral charge balance.

Forces Between Charges

Opposite charges attract. Like charges repel. At an elemental level, like charges always repel (electrons repel electrons, protons repel protons), but for macroscopic objects, non-symmetric charge distribution can result in an overall attraction between two objects that carry the same type of overall charge (positive or negative). Non-symmetrical charge distribution always results in an attraction between a charged object and an electrically neutral (overall) object.

Charging

All charging processes involve the transfer of electrons from one object to another. In order for an object to become positively charged, it must lose some of its electrons. In order for an object to become negatively charged, it must acquire more electrons.

Charging by rubbing: When two initially neutral non-conducting objects are rubbed together, one of them will generally bind electrons more strongly and take electrons from the other material. The law of conservation of charge requires that the total amount of electrons be conserved. That is, electrons only move from one object to another, but no electrons are created or destroyed. Overall, the two objects together still have a net charge of zero.

Charging by contact: When a charged object is touched to a neutral or less charged object, repulsive forces between the like charges result in some of the charge transferring to the less charged object so the like charges will be further apart. This effect is much larger for conducting objects.

Charging by induction: The protons and electrons inside any object respond to electric forces of attraction or repulsion. When an object is placed near a charged object, the charged object will exert forces on the protons and electrons inside the other object, forcing them to move apart. One side of the object will become more positive than it was initially. The other side will become more negative, as electrons migrate internally. This condition is called **polarization**, a word that refers to the object having "poles," or opposite sides with different electrical states, even though the object as a whole may still be charge neutral. If a conductor is touched to the polarized object, some of the charge will transfer to the conductor. If the conductor is then removed, the object will carry a net charge different from its initial charge. For example, consider the sequence in Figure 1. Figure 1a shows an isolated neutral conductor. In Figure 1b, a negatively charged object has been placed near the neutral conductor, which is now polarized. Another conductor (elliptical) is shown but is not yet involved. In Figure 1c, the elliptical conductor touches the polarized conductor and some of negative charge transfers to the elliptical conductor due to repulsive force from the negative charged object. In Figure 1d, the elliptical conductor is removed, taking some negative charge with it and leaving a net positive charge on the round conductor.

In Figure 1(e), the negatively charged object has been removed. The charge on the round conductor re-distributes, but the overall charge on the conductor is now positive. Note that the polarization of the elliptical conductor was ignored.

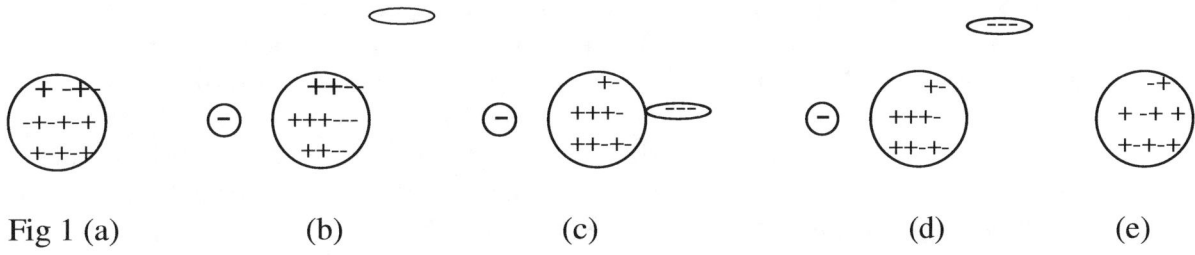

Fig 1 (a)   (b)   (c)   (d)   (e)

Equipment

1	Basic Electrometer	ES-9078
1	Charge Producers and Proof Plane	ES-9057C
1	Faraday Ice Pail and Shield	ES-9042A
1	Conductive Spheres, 13 cm	ES-9059C
1	Conductive Shapes	ES-9061
1	Electrostatics Voltage Source	ES-9077
1	PASCO Capstone	UI-5400

Figure 2: Charging Setup

Setup A

1. Connect the Electrometer to the Faraday Ice Pail as shown in Figure 2. Connect the red alligator clip to the inner conductor and the ground (black alligator) to the outside conductor. The Electrometer measures the voltage difference between small charges without affecting the charges. The "ice pail" is the inner conducting mesh cylinder. When a charge Q is placed inside the ice pail, it becomes polarized with a charge almost equal to Q moving to the outside of the inner cylinder. The voltage between the inner and outer cylinders is directly proportional to the charge on the outside of the inner conductor. Therefore the charge inside the ice pail can be directly measured.
2. Turn on the Electrometer and set the Range to 100V.
3. Ground the Ice Pail by touching a finger or wire to both the inner and outer cylinder at the same time. Remove the finger or wire from the inner cylinder first then lift off of the outer cylinder. To "ground" means to remove most of the excess charge from the system.
4. Press the "ZERO" button on the Electrometer. This sets the Electrometer to read zero even if there is a charge on the Ice Pail. This is OK since we really are only interested in changes in the charge.
5. You may need to repeat the grounding and/or zeroing of the Ice Pail often during the experiment. It is very easy to transfer charge to the ice pail by touching it or even getting too close to it with a charged object. It may even acquire a charge sitting on the table for a while. To see how sensitive the system is, stick a finger down the axis of the inner cylinder (without touching the cylinder.) Now rub your fingers through your hair, or on your shirt, or shuffle your shoes on the floor and try sticking your finger back into the Ice Pail. See any difference? What happens if you touch the Ice Pail? If you checked this reground the system.
6. One at a time, insert the disks on the Charge Producer wands (one with a dark plastic disk and the other with a white leather disk) into the inner basket. If the disks are uncharged, the needle in the electrometer will stay at zero. If the needle moves, then there is residual charge in the wands. To remove the residual charge, touch the wands to the outer basket of the grounded ice pail. Sometimes the residual charge is hard to remove. You can breathe on the disks, then touch the disk to the outer basket again to remove the remaining charge. The moisture from your breath helps remove the charges. It is difficult to remove all the charge since the disks are non-conducting. If the voltage change is less than 1V for each wand in the ice pail, continue onto the procedure.

Last/First Name (print): _____

PHYS 1440-Section _____ Username ID (xxx9999): _____

Procedure A1: Charging by Rubbing Objects Together

Important Note: This procedure needs to be performed quickly to avoid charge migration, therefore read procedure first before beginning.

1. Ground the ice pail.
2. Zero the electrometer.
3. Make sure there is no charge on the charge producing wands (see step 6 under Setup A).
4. Insert both wands into pail.
5. Zero the electrometer.
6. Record "initial reading".
7. Rub wands together quickly for 5 seconds.
8. Record "after rubbing".
9. Separate the wands.
10. Record "after separation."
11. Take dark wand out of pail.
12. Record "dark out."
13. Insert dark wand back into pail, careful not to touch pail or white wand.
14. Record "both in."
15. Take the white wand out of the pail.
16. Record "white out."
17. Insert white wand back into pail, careful not to touch pail or dark wand.
18. Record "both in final."
19. Remove both wands from the pail.
20. Record "both out."
21. Repeat the procedure three times, then check with your TA to be certain you have reasonable readings.

Note: The Electrometer Voltage table lists the electrometer voltage for each of the cases described above for each of the three runs.

Table A1: Electrometer Voltage

Line #	Recording Condition	Trial 1	Trial 2	Trial 3
1	Initial reading			
2	After rubbing			
3	After separation			
4	Dark out			
5	Both in			
6	White out			
7	Both in final			
8	Both out			
9	Total of row 4 + row 6			

Last/First Name (print): _____

PHYS 1440-Section _____ Username ID (xxx9999): _____

Procedure A2: Charging by Contact

1. Ground the ice pail and zero the electrometer.
2. Record the "zero" reading in line 1 of the Charge by Contact table.
3. Briskly rub the white and dark charge producers together outside of the pail to charge them for 5 seconds.
4. Insert the white charge producer into the inner basket of the ice pail.
5. Record the "initial" voltage on the electrometer in line 2 of the Charge by Contact table.
6. Touch the wall of the basket with the white disk for 1 second.
7. Record the "after touch" voltage on the electrometer in line 3 of the Charge by Contact table.
8. Remove the white disk.
9. Record the "Disk Out" voltage on the electrometer in line 4 of the Charge by Contact table.
10. Repeat steps 1-9 using the dark wand.

Table A2: Charge by Contact Table

Line #	Recording Condition	White Disk			Black Disk		
		Trial 1	Trial 2	Trial 3	Trial 1	Trial 2	Trial 3
1	Zero reading						
2	Initial voltage with disk in						
3	After touch						
4	Disk out						

Procedure A3: Charging by Induction

1. Ground the ice pail and zero the electrometer.
2. Record the "zero" reading in line 1 of the Charge by Induction table.
3. Rub the wands together to charge them for 5 seconds.
4. Without letting the wand touch the ice pail, insert the white wand into the lower half of the inner basket.
5. Record the "initial" voltage on the electrometer in line 2 of the Charge by Induction table.
6. Ground the ice pail by bridging between the outer and the inner cylinder with your finger and then removing your finger. The white wand remains in the basket for this step.
7. Record the "after grounding" voltage on the electrometer in line 3 of the Charge by Induction table.
8. Remove the wand from the ice pail.
9. Record the "wand out" voltage on the electrometer in line 4 of the Charge by Induction table.
10. Repeat steps 1-9 using the dark wand.

Table A3: Charge by Induction

Line #	Recording Condition	White Disk			Dark Disk		
		Trial 1	Trial 2	Trial 3	Trial 1	Trial 2	Trial 3
1	Zero reading						
2	Initial voltage with disk in						
3	After grounding						
4	Disk out						

Last/First Name (print): _____

PHYS 1440-Section _____ Username ID (xxx9999): _____

Analysis A1:

1. What can you immediately conclude about the charges on the white wand and the dark wand based on the signs of the voltages in Table A1?

2. Note that lines 3, 5, and 7 are exactly the same system so should have the same voltage. Any difference implies that there has been some charge lost or gained. Line 8 should be zero unless some charge has been transferred. An acceptable uncertainty is 2-3 V. Remark whether or not the numbers agree within uncertainty and if charge appears to have been conserved or not.

3. Recall that the voltage is directly proportional to the charge. Thus if a voltage of 8 V implies 8 units of charge, a voltage of 12 V implies 12 units of charge. What do lines 1-3 imply?

4. Compare line 1 to line 9. What does this imply?

Analysis A2:

5. What can you conclude from the data in Table A2 lines 2-4?

Analysis A3:

6. Explain what is happening when you induce a charge on the inner mesh cylinder. That is, explain the data in lines 3 & 4 of Table A3. Why does the sign change between "initial" and "disk out"?

Last/First Name (print): _____

PHYS 1440-Section _____ Username ID (xxx9999): _____

Theory B

Charge Distribution on a Conducting Surface
A charge will tend to concentrate at places on a conductor where the surface is more sharply curved. This happens because charges do not interact as strongly with other charges that are "over the horizon" since the electric field lines cannot pass through a conductor.

Charge Inside a Spherical Conducting Shell
The force between two point charge particles obey Coulomb's law. q_1 and q_2 are the charge of each particle, r is the distance between them, and ϵ_0 is the permittivity of free space.

$$\textbf{Coulomb's Law: } F = \frac{1}{4\pi\epsilon_0}\frac{q_1 q_2}{r^2}$$

Because the force between two point charges obeys an *inverse square law* (i.e. falls off like $1/r^2$), then all the charge on a conducting spherical shell must lie on the outside of the shell and the inside of the shell must be uncharged. Although we check this crudely here, it is possible to verify this with great precision.

Setup B

1. Plug in the Electrostatics Voltage Source and turn it on.
2. Attach a black lead from the Common (Com) jack to a convenient ground like a water pipe or the silver outside connectors of the #2 or #3 outputs on the 850 Universal Interface.
3. Attach the red lead to the +3000 V terminal and leave the spade connector on one end of the red lead unattached. P.S. you cannot get shocked by this power supply since it can only produce a tiny current.
4. Ground the ice pail (the inner cylindrical wire mesh) with your finger as before.
5. Turn on the Electrometer, set it to the 30V range, and zero the electrometer. If necessary, set the electrometer reading to 100V range.

Procedure B1: Measuring the Charge at Different Locations on the Outer surface of a Spherical Conductor

The "proof plane" is the metal disk (silver looking) on one of the black wands. It is used for sampling charge on a conductor by touching it to the surface of the conductor. The amount of charge that transfers to the proof plane is proportional to the surface charge density at the point of contact.

1. Make sure there is no charge in the proof plane by inserting it into the inner mesh cylinder. The electrometer should read zero. If there is charge on the proof plane, ground it by touching it to any ground or the outer mesh cylinder.
2. Place the spherical conductor (no hole) well away from other objects (including people) to prevent polarization effects. Nothing should be closer to the sphere than the base on which it sits. The Faraday Ice Pail should also be well away from the sphere.

You should perform the following steps as rapidly as possible to prevent significant discharge of the sphere though the air.

3. Briefly touch the free end of the red lead (the spade lug) to the spherical conductor (without a hole). This will charge the conductor to an electric potential of 3000V.
4. Touch the sphere at point A (see Figure 3) with the proof plane keeping your hand as far away as possible.
5. Put the proof plane in the inner cylinder of the "ice pail" and record the voltage from the electrometer on Figure 3 at point A. Do not ground the proof plane so the total charge on the sphere and proof plane remains constant.
6. Repeat for points B & C.

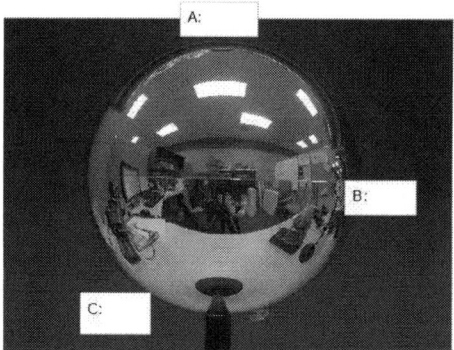

Figure 3: Charge Distribution on a Sphere

Table B1: Position vs. Charge on Sphere

Position	Electrometer Reading		
	Trial 1	Trial 2	Trial 3
A			
B			
C			

Last/First Name (print): _____

PHYS 1440-Section _____ Username ID (xxx9999): _____

Procedure B2: Charging the Spherical Conductor to Different Potentials

1. **Same setup as in Procedure B1.**
2. Ground the ice pail.
3. Set the electrometer to the 10 V range and zero the electrometer.
4. Make sure there is no charge in the proof plane. The electrometer should read zero. Repeat this anytime the electrometer does not read zero with nothing in the ice pail.
5. Touch the other end briefly to the surface of the conducting sphere with one end of the red lead in the +1000 V terminal of the Electrostatics Voltage Source.
6. Touch the top of the sphere with the proof plane.
7. Insert the proof plane into the lower half of the inner basket of the ice pail without letting it touch the basket.
8. Record the electrometer voltage in the second column of the Table B2.
9. Ground the proof plane. Make sure there is no residual charge in the proof plane before continuing.
10. Repeat the measurement for the +2000 V and +3000 V outputs (still on the 10 V range). If the reading is 9.9 V, you have exceeded the range. In that case, change the electrometer to the 30 V range, ground the ice pail, zero the electrometer and proceed.

Table B2: Voltage Applied vs Charge on Sphere

Sphere Potential	Electrometer Reading at Position A		
	Trial 1	Trial 2	Trial 3
+1000 V			
+2000 V			
+3000 V			

Procedure B3: Measuring the Charge on the Outer Surface of a Non-Spherical Conductor

1. Same setup as Procedure B1 except replace the sphere with the non-spherical conductor. Use the +3000 V terminal on the Voltage Source. Electrometer on the 30 V range.
2. Ground the ice pail, zero the electrometer and make sure there is no charge in the proof plane. The electrometer should read zero.
3. Briefly connect the oblong-shaped conductor to the +3000V outlet of the Electrostatics Voltage Source.
4. Touch the large-radius end (point A on Figure 4) of the oblong shaped conductor with the proof plane.
5. Without letting it touch the basket, insert the proof plane into the lower half of the inner basket of the ice pail. Record the reading from the electrometer on Figure 4 at point A
6. Repeat the process at points B & C. Do not ground the proof plane so the total charge remains constant.

Figure 4: Non-spherical Charge Distribution

Table B3: Position vs. Charge on Non-Spherical Conductor

Position	Electrometer Reading		
	Trial 1	Trial 2	Trial 3
A			
B			
C			

Last/First Name (print): _____

PHYS 1440-Section _____ Username ID (xxx9999): _____

Procedure B4: Measuring the Charge on the Inner Surface of a Spherical Conductor

1. Same setup as Procedure B3 except replace the oblong conductor with the sphere with a hole on top.
2. Ground the ice pail, zero the electrometer and make sure there is no charge on the proof-plane wand (silver). Set the electrometer to the 30 V range.
3. Briefly touch the end of the red lead (attached to the +3000 V terminal) to the conducting sphere.
4. Insert the uncharged proof plane into the charged sphere and touch the inner wall near the bottom, keeping the proof plane as far as possible from the edges of the hole. Little charges on the handle are not important. Considerable care is required to insert the disk into the hole without touching the sphere. There is a risk of charge transfer even if the disk comes too close to the edge of the hole without touching.
5. Remove the proof plane from the hole. If the proof plane touches the edge, start over.
6. Insert the proof plane into the lower half of the inner basket of the ice pail without letting it touch the basket.
7. Record the electrometer reading in line 1 of Table B4.
8. Insert the uncharged proof plane into the charged sphere and touch the inner wall near the hole keeping the proof plane as far as possible from the edges of the hole
9. Remove the proof plane from the hole. If the proof plane touches the edge, start over.
10. Insert the proof plane into the lower half of the inner basket of the ice pail without letting it touch the basket.
11. Record the electrometer reading in the second line 2 of Table B4.
12. Touch the proof plane to the outside of the sphere (anywhere) and repeat step 10, recording the value from the electrometer on line 3.
13. Insert the proof plane in the hole and touch the inner wall near the bottom **without grounding it.**
14. Repeat step 10, recording the value on line 4.
15. Touch the proof plane to the outside of the sphere (anywhere).
16. Repeat step 10, recording the value from the electrometer on line 5. If lines 3 and 5 are nearly equal and around 10 V, the data is OK. Variations of +/- 1 V are not significant.

Table B4: Sphere Voltages

Line #	Position	Electrometer Reading		
		Trial 1	Trial 2	Trial 3
1	Charge on Al wand after touching the inner bottom wall			
2	Charge on Al wand after touching the inside, near the opening			
3	Charge on Al wand after touching outside surface			
4	Charge on Al wand after touching the inner bottom wall again			
5	Charge on Al wand after touching the outside surface again			

Last/First Name (print): _____

PHYS 1440-Section _____ Username ID (xxx9999): _____

Analysis B1:

Recall that the electrometer voltages are not the voltages on the sphere (the sphere is at a uniform 3000 V). The recorded data is the voltage between the inner and outer mesh cylinders when the proof plane is inside the inner mesh cylinder. This voltage is directly proportional to the charge on the proof plane. Using Table B1, what can you conclude about the distribution of charge on the sphere?

Analysis B2:

Recall that the electrometer voltages are the voltages between the inner and outer mesh cylinders with the proof plane inside the inner mesh cylinder. This voltage is directly proportional to the charge on the proof plane. Using Table B2, what can you conclude about the electric potential (voltage) of the sphere vs the charge on the sphere?

Analysis B3:

Using Table B3, what can you conclude about the distribution of the charge on a non-spherical conductor?

Analysis B4:

1. What does the data in Table B4 tell you about the distribution of charge on a hollow spherical shell?

2. Explain what happened in line 4 of Table B4 (step 13 of B4 procedure).

1440 Questionnaire: Experiment 2: Electrostatic Charging

Do not put your name on this page. Hand in this page separately.

1. TA name(s):

2. What one thing did you like best about this laboratory?

3. What one thing did you like least about this laboratory?

4. What one thing would you change in this laboratory?

5. What one thing would you leave the same?

Additional Comments?

Last/First Name (print): _____

PHYS 1440-Section _____ Username ID (xxx9999): _____

Experiment 3:
Electric Field Plotting

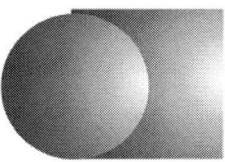

Pre-lab

1. Define electric field, and show (in equation form) how it is related to force.

2. Can electric field lines cross each other? Why or why not?

3. How much work is done to move a charge 1.5 meters using 10 Newton of force across an <u>equipotential</u> surface?

4. Briefly describe the relationship between an equipotential surface and an electric field, and use this to explain why we will plot equipotential lines.

5. You are using a voltmeter to measure the voltage at a particular point on an equipotential line, and the voltmeter reads 2 Volts. If you instead measure a different point on the same equipotential line, is it possible to know what the voltmeter should read? If so, what should it read?

Last/First Name (print): _____

Experiment 3:
Electric Field Plotting

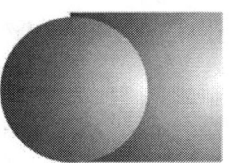

Introduction

The purpose of this <u>qualitative</u> activity is to introduce the students to the concept of an electric field and to make the idea more concrete by examining a number of examples. Several rules about the electric field are verified. At the end of the activity, the student should be able to sketch the electric field around a simple charge distribution. The idea of equipotential surfaces is also developed.

Theory

The electric field is way to visualize the interaction between charged objects. It is assumed that the electric field permeates the space around a charged object and is responsible for the force that the object exerts on other charges. The electric field (strength) is defined as the force per unit charge on a vanishingly small test charge q.

$$\mathbf{E} = \mathbf{F}/q$$

Note that this is a vector relationship where **E** and **F** must lie in the same (or opposite if q is negative) direction.

To help visualize the field we use the concept of electric field lines where, by convention, the lines begin on positive charges and terminate on negative charges. These lines must satisfy two conditions.
   1. **E** must be tangent to the lines at all places.
   2. The magnitude of **E** is directly proportional to the density of the lines.

The second requirement is only true in a three dimensional space. The patterns in this activity will be essential two dimensional (due to the conductive paper), however it will still be true that where the lines are close together, the field is strong.

There are a number of rules about the field lines that help us draw them for the static case:
1. The lines should have the same symmetry as does the charge distribution.
2. The number of lines from a charge **q** should be proportional to **q**. Note: due to the way the power supply works, the charge due to the positive terminal is equal and opposite of the charge due to the negative terminal.
3. The point charge dominates (due to $1/r^2$) and the field is that of an isolated point charge (fully symmetric) near a point charge.
4. The lines cannot stop in space. They begin on a + charge and end on a – charge. (Otherwise the force does not fall off as $1/r^2$.)
5. The field lines cannot cross (otherwise it violates the conservation of energy.)
6. The field lines cannot pass through a conductor (otherwise charge moves.)
7. The field lines must be perpendicular to the surface of a conductor (otherwise charge moves.)
8. The lines must be perpendicular to any equipotential surface (see below).

An equipotential surface is a surface upon which no work is required to move a charge. How can we move a charge in a region where there is a force on it due to the electric field and yet do no work? The answer lies in the fact that the work, W, done is the scalar product of force, **F**, and displacement, **d**, so depends on the angle.

$$W = Fd \cos \theta$$

If θ is 90^0, the work done is zero. For example, a satellite in a circular orbit around the Earth does no work against gravity so does not need any additional energy (i.e. firing its engine) to orbit. So an equipotential surface must be perpendicular to the force and thus <u>an equipotential surface must be perpendicular to the electric field (strength) and field lines</u> (see Figure 1). The electric field is difficult to measure directly, but it is very easy to find the equipotential surfaces using a voltmeter (we'll see how a voltmeter works later in the course). For example, all the points where the voltage is 4 volts form an equipotential surface. In the case of our two dimensional pattern, the equipotential "surfaces" will be lines.

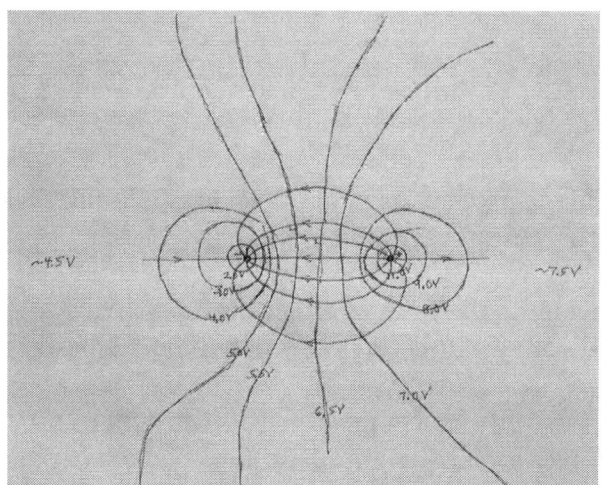

Figure 1: Dipole Field

Equipment

1	Field Mapper Kit	PK-9023
	Replacement Supplies:	
	Conductive Ink Pen	PK-9031B
	Conductive Paper w grid	PK-9025
	Conductive Paper (no grid)	PK-9026
1	Voltage Sensor	UI-5100
1	Short Patch Cords	SE-7123
1	PASCO Capstone	UI-5400
1	850 Universal Interface	UI-5000

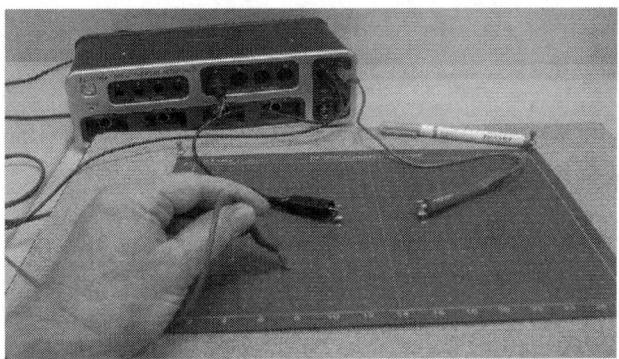

Figure 2: Setup

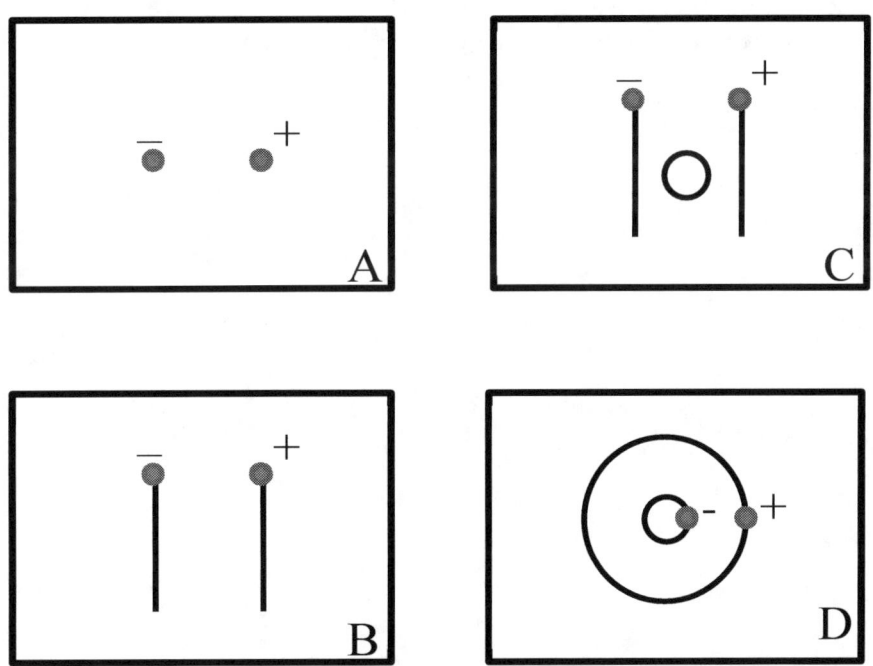

Figure 3: Patterns

49

Setup

1. Prepare the patterns in Figure 3 using the conductive pen and conductive paper. The patterns must dry completely before use. The letters are for reference only and should not be drawn on the paper. The + and – are to show where to connect the electrodes and should not be drawn on the paper. The circles in patterns C are free floating and should not be connected to the power supply.
2. Attach the conductive paper with the pattern on it to the cork board with push pins at the four corners. The paper is not very conductive, but must carry a little current for the voltage sensor to work. The fact that the paper carries a small current actually affects the electric field somewhat and near the edges of the paper the effects may be noticeable.
3. Attach a red patch cord to the red terminal of Output 1.
4. Attach a black patch cord to the black terminal.
5. Attach an alligator clip to the other end of each patch cord and clip each to a silver push pin (see Figure 2).
6. Insert the push pins into the appropriate electrodes on the pattern drawn on the conductive paper. *It is important that you do not bump the electrode push pins during the experiment* since doing so may change the voltages on the paper. This is not a disaster since the shapes of the equipotential surfaces will not be affected, but it is frustrating to be tracing a 4.0 V surface that suddenly becomes a 3.0 V surface.
7. Click on Signal Generator at the left of the screen.
8. Set Output 1 for a 15 V DC signal
9. Click On so it will turn on any time you click Record.
10. Click the Signal generator again to close it.
11. Insert the Voltage Sensor in Analog Input A on the 850 Universal Interface.
12. Attach the black lead to the black terminal on Output 1 of the 850.

Procedure A: Plotting Equipotential Lines

1. Check for good connections by first clicking Record and touching the red lead of the voltage sensor to the paper near (not touching) the negative electrode. The voltage reading (upper right of screen) should be 2-3 V. The value near the positive electrode should be at least 12 V. For the conductors drawn, make sure the voltage is the same everywhere on the circle or line. If not there are breaks in the line. Get a new pattern.
2. Draw your pattern onto the white grid paper to scale.
3. Trace the 4.0 V equipotential. To do this, locate a place on the paper near the positive electrode where the voltage is 4.0 V. Try to keep the probe at the same angle (Figure 4) and use the same pressure for each reading (the voltage is pressure sensitive). Record the point on your white grid paper and write 4 V next to it. *Hint: near a point electrode, the 4.0 V equipotential should be roughly circular. For other electrodes, the 4.0 V equipotential should roughly be parallel to the electrode.* ***Important: This experiment is not precise. 4.0 V will be within a few tenths of a volt. Do not waste time trying to find the exact spot! This process is meant to be quick.***
4. Locate 5 more 4.0 V points to make an equipotential surface. Mark these points on the white grid paper.
5. This process will be repeated again with a different voltage to create another equipotential surface. <u>Find 6 more equipotential surfaces</u> by picking new voltages. Record the points onto the white grid paper and write the voltage next to each point.
6. Click Stop when you have ten equipotential surfaces.

Figure 4: Probe Position

7. Connect the points that are at the same voltage. Draw smooth curves if they don't go exactly through each data point (Figure 5). Everything on this line is an equipotential surface.

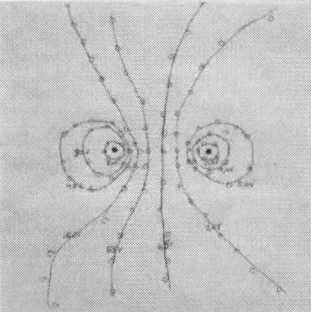

Figure 5: Equipotential Surfaces

Procedure B: Plotting Electric Field Lines

1. Tape the two leads of the voltmeter together for this procedure (see Figure 7). The technique is to use the voltmeter leads to find the direction from an electrode that follows the path of greatest potential difference from point-to-point.
   NOTE: Do not attempt to make measurements by placing the leads on the grid marks on the conductive paper. Touch the voltmeter leads only on the solid black areas of the paper. It may be necessary to use higher voltmeter sensitivity for this measurement than was used in measuring equipotentials.
2. Place the voltmeter lead connected to ground near one of the dipoles to plot the field lines on the conductive paper.
3. Place the other voltmeter lead on the paper and note the voltmeter reading.
4. Pivot the lead to several new positions while keeping the ground lead stationary (see Figure 7).
5. Note the voltmeter readings as you touch the lead at each new spot on the paper.
6. Draw an arrow on the paper from the ground lead to the other lead (see Figure 8) when the potential is the highest,.
7. Move the ground lead to the tip (head) of the arrow.
8. Repeat the action of pivoting and touching with the front lead until the potential reading in a given direction is highest.
9. Draw a new arrow.
10. Repeat the action of putting the ground lead at the tip (head) of each new arrow and finding the direction in which the potential difference is highest. Eventually, the arrows drawn in this manner will form a field line.
11. Return to the dipole and select a new point at which to place the voltmeter's ground lead.
12. Probe with the other lead until the direction of highest potential difference is found.
13. Draw an arrow from the ground lead to the other lead
14. Repeat the process until a new field line is drawn (see Figure 9).
15. Continue selecting new points and drawing field lines around the original dipole (see Figure 6).

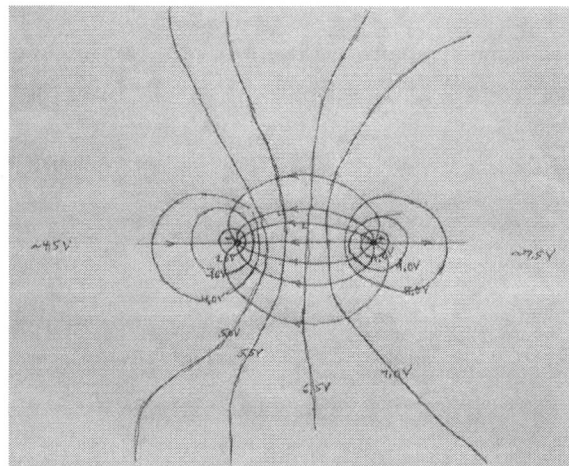

Figure 6: E Field Lines

Note: equipotential lines near the edge of the paper will be distorted. The system is not quite static since some current is required for the voltage sensor to function. The E field lines show the path that charge (current) follows. Current cannot flow from the conducting paper into the air, so the field lines must either run parallel to the edge or avoid it. Places to the left and right of the pattern show an almost constant voltage, so in these places the electric field must be nearly zero. At the top and bottom of the pattern, the E field lines run parallel to the edge and the equipotentials must be perpendicular to the edge. This modification of the field is produced by surface charge on the edge of the paper.

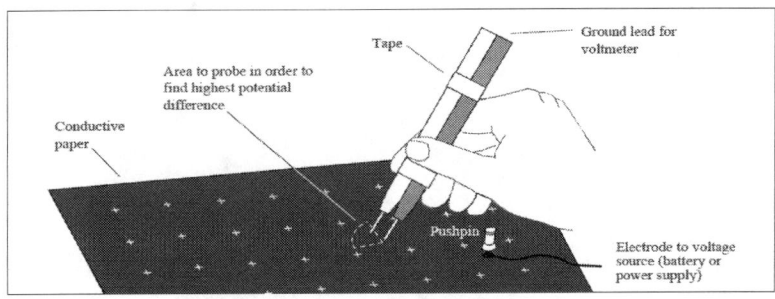

Figure 7: E Field plot setup

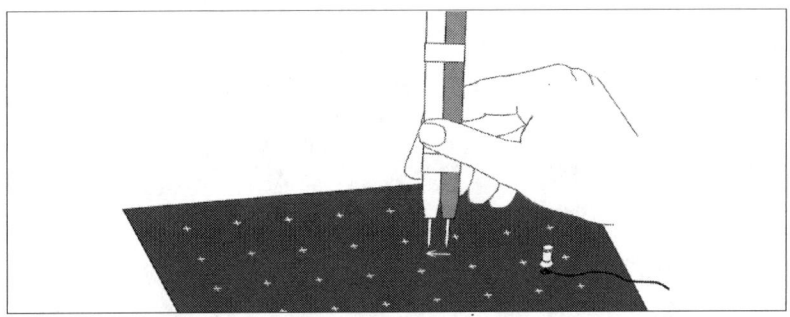

Figure 8: Probe position showing highest potential

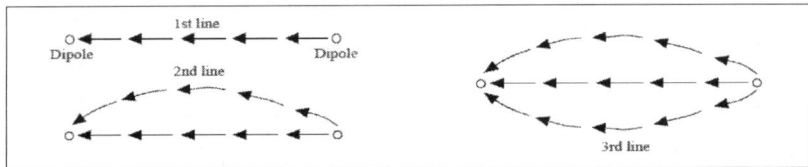

Figure 9: Example of three field lines between unlike dipoles

Last/First Name (print): _____

PHYS 1440-Section _____ Username ID (xxx9999): _____

Analysis:

Pattern A: Attractive Dipole

1. What is the relation between the direction of electric field and equipotential line at the same point? (A geometrical relation is desired.)

2. What effect does the finite size of the black paper have on the field?

Pattern B: Parallel Plate Capacitor

1. What is the field outside the capacitor plates?

2. How does the ratio of the plate length (l) versus separation (d) affect the fringing effect at the edges of the plates?

Pattern C: Floating Electrode

1. How does the circular electrode distort the field?

2. What is the potential of the circular electrode? Of the area inside the electrode?

3. What effect would moving the circular electrode have?

Pattern D: Coaxial Cable

1. What is the field inside the inner conductor, between the two conductors, and outside the outer conductor?

Last/First Name (print): _____

PHYS 1440-Section _____ Username ID (xxx9999): _____

Conclusions:

Discuss any interesting points about your pattern. As appropriate, answer each of the questions that follow. It is interesting to note that pattern A is the attractive dipole that show in most textbooks. Pattern B is a parallel plate capacitor, pattern C is a floating electrode between a parallel plate capacitor, and pattern D is a coaxial cable.

1. Draw the distribution of charge for the conductor(s) in your pattern. Remember that electric field lines begin on positive charges and terminate on negative charges. A higher field line density implies a larger charge, and a lower field line density implies a smaller charge. Write a + where each line begins and a − where each line terminates.

2. Does your pattern have the same symmetry as the charge distribution?

3. Do the field lines cross each other, should they cross?

4. Are the field lines perpendicular to the surface of your conductor?

5. Where in the pattern is the field strongest/weakest? Note that where the equipotential surfaces are close together, the electric field must be strong (why?)

6. Is there any evidence of charge polarization in your pattern?

1440 Questionnaire: Experiment 3: Electric Field Plotting

Do not put your name on this page. Hand in this page separately.

1. TA name(s):

2. What one thing did you like best about this laboratory?

3. What one thing did you like least about this laboratory?

4. What one thing would you change in this laboratory?

5. What one thing would you leave the same?

Additional Comments?

Last/ First Name (print): _____

PHYS 1440-Section _____ Username ID (xxx9999): _____

Experiment 4:

Ohm's Law

Pre-lab

1. Define the relationship between voltages and current in metals (what law is associated with this?)

2. Do all devices have to obey Ohm's Law?

3. Does resistance in a circuit remain constant if the temperature changes?

4. Can a diode conduct electricity in a reverse bias orientation at low voltages?

Last/First Name (print): _____

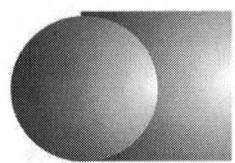

Experiment 4:
Ohm's Law

Introduction

The purpose of this experiment is to verify Ohm's law for commercially manufactured carbon resistors and to examine current and voltage relationships in a light bulb and a diode. In a lightbulb, as the current increases the energy dissipated in the bulb (or any other resistor) increases as well. This results in the increased temperature of the bulb and a change in the resistance. One can observe the corresponding changes in the current and voltage characteristic slope. In a diode the relationship between current and voltage will be analyzed for 2 different orientations (forward and reverse bias).

Theory

The electric current that flows through most conductors of electricity (example: metals) is directly proportional to the voltage applied across them (provided the temperature remains constant):

$$V \propto I,$$

This is referred to as Ohm's Law. It is convenient to define a proportionality constant called the resistance (unit: Ohm $[\Omega]$ = V/A) such that

$$V = IR. \qquad \textbf{Equation 1}$$

The ratio of voltage to current is called resistance. An "ohmic" material has this ratio (resistance) constant. A resistor generally means a device that obeys Ohm's Law (many devices do not) and has a resistance R.

An example of a device that does not follow Ohm's Law is a diode. A diode is a common device that only allows electricity to flow in one direction. Diodes are used to convert AC to DC, build logic gates, and limit voltages.

A diode (appearance shown in Figure 4) is an electronic device, where a junction between two semiconductors (p and n-type) can conduct current at forward bias (Figure 5).

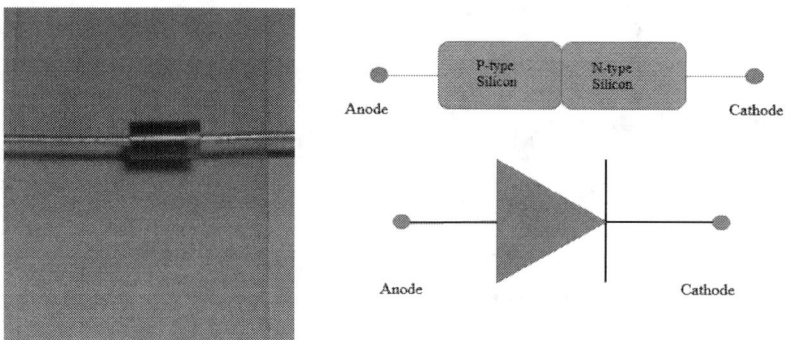

Figure 4 **Figure 5**

- Forward bias is when the positive side of the voltage source is on the anode (p-type) and the negative side is on the cathode (n-type). In this case, the diode becomes a conductor and allows current to flow.
- Reverse bias is when you reverse the voltage direction, applying positive bias to the cathode and negative bias to the anode. In this case, current does not flow and the diode becomes an insulator. A diode cannot conduct at reverse bias unless breakdown occurs at very high voltages
- Because of these characteristics, the diode is considered to be an electronic switch. It is used in rectifier circuits (converting AC to DC).

Equipment

1	AC/DC Electronics Laboratory	EM-8656
1	Short Patch Cords (set of 8)	SE-7123
1	850 Universal Interface	UI-5000
1	PASCO Capstone	UI-5400

Setup

1. Connect red and black patch cables from the signal generator to the board as shown in figure 1.
2. Open the signal generator at the left of the screen. Set Output 1 to a triangle waveform with a frequency of 0.1 Hz and an amplitude of 2.5 V. Select auto. Click the signal generator again to close the panel.

Procedure: Verifying Ohm's Law for a Resistor

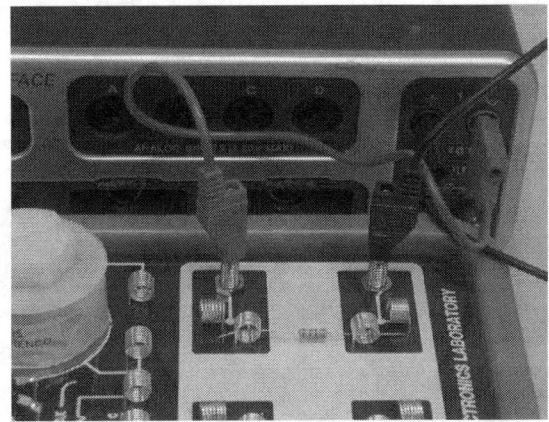
Figure 1: Ohm's Law Setup

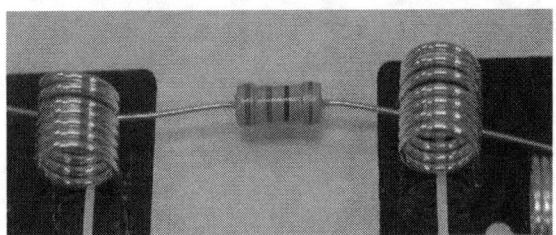
Figure 2: Resistor Color Code

1. Setup the circuit shown in Figure 1 using the 330 Ω (orange-orange-brown-gold) resistor between the spring clips. The gold band on the resistor means it is accurate to within 5% (330 +/- 17 Ω).
2. Click RECORD. Collect data for 10 seconds (one full cycle) and click STOP. The graph (Output Voltage vs. Output Current) should show a straight line (with some noise.)
3. Click on Data Summary at the left of the page. Double click on any Run #1 and re-label it "330 Run". Click Data Summary again to close it.
4. Replace the 330 Ω resistor with a 560 Ω (green-blue-brown-gold) resistor.
5. Click RECORD. Collect data for 10 seconds and click STOP. Re-label this run "560 Run".

Procedure: Current-Voltage Relationship for a Light Bulb

1. Remove the 560 Ω resistor and attach a light bulb to the input terminals using long wires as jumpers as shown in Figure 3.
2. Click RECORD. Collect data for 30 seconds and click STOP. There should be a kink in the graph. Observe what the bulb is doing as the kink occurs.

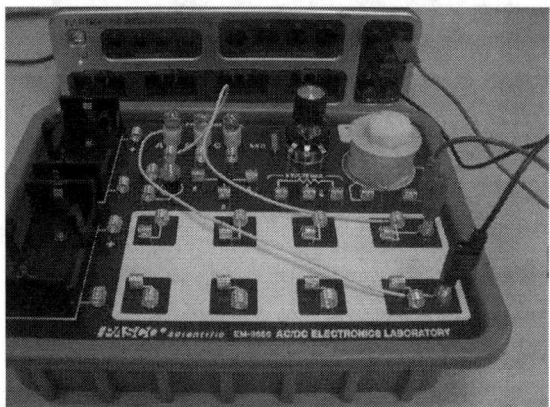

Figure 3: Light Bulb Setup

Procedure: Current-Voltage Relationship for a Diode

1. Click open the Signal Generator and reduce the voltage to 1.0 V.
2. Remove the jumper wires and place the diode in the clips as in Figure 4. The diode looks like a small black cylinder with one grey end. Arrange the diode so the grey end is toward the red lead from the 850 Interface.
2. Click RECORD. Collect data for 10 seconds and then click STOP. Re-label this run as "forward bias".
3. Reverse the orientation of the diode so the grey end is towards the black lead.
4. Click RECORD. Collect data for 10 seconds and click STOP. Re-label this run as "reverse bias".

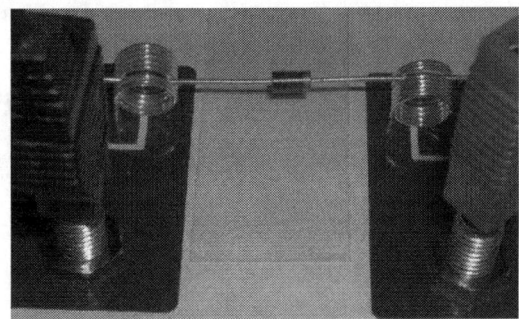

Figure 4: Diode Up Setup

\

Analysis: Verifying Ohm's Law for a Resistor

Random noise generated in the 850 Universal Interface limits the precision of the current reading to about +/- 0.1 mA. In addition, there is a zero error in the current reading of up to 5 mA. This affects the intercepts but does not change the slope of the line.

1. Click once on the multicolored Run Select icon in the graph toolbar to show more than one data set.
2. Click on the black triangle by the Run Select icon and select the "330 Run" and the "560 Run". Both should show on the graph.
3. Click on the Scale to Fit icon on graph toolbar. Sketch the data below in Graph 1.
4. Select the "330 Run" data by clicking the "330 Run" data on the graph.
5. Click on the black triangle by the Curve Fit icon on the graph toolbar and apply a linear fit.
6. Repeat for the "560 Run" data. You may need to drag one of the linear boxes to see both of them at the same time.

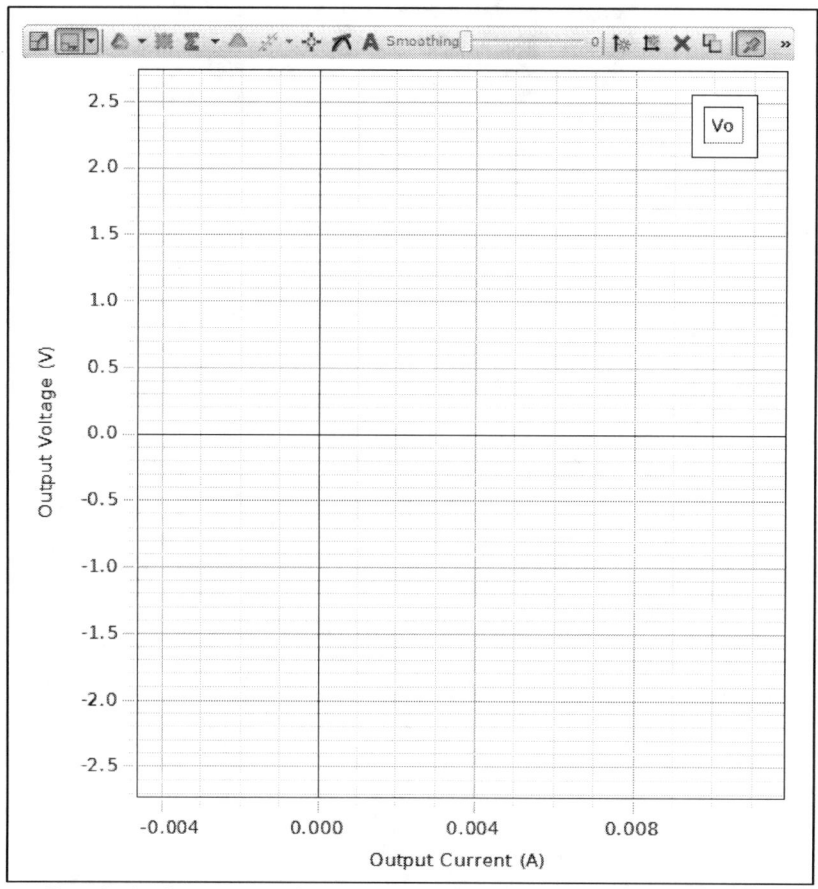

Graph 1: Output Voltage vs. Output Current for a Resistor

Analysis: Light Bulb

1. Click on the black triangle by the Run Select icon and select the "Light Bulb" run.
2. Click the Scale to Fit icon.
3. Click once on the multicolored Run Select icon to show more than one data set.
4. Click on the black triangle by the Run Select icon and select the "Light Bulb" data. Sketch the data in Graph 2.

   Wait a minute! Tungsten is a metal and the metals are supposed to obey Ohm's Law. What is going on?

5. Click on the Selection icon and drag the Selection box handles to select the voltage data between -1 V and +1 V and the current data between -0.1 A and +0.1 A.
6. Click the Scale to Fit icon.
7. Click on the black triangle by the Curve Fit icon and apply a linear fit.

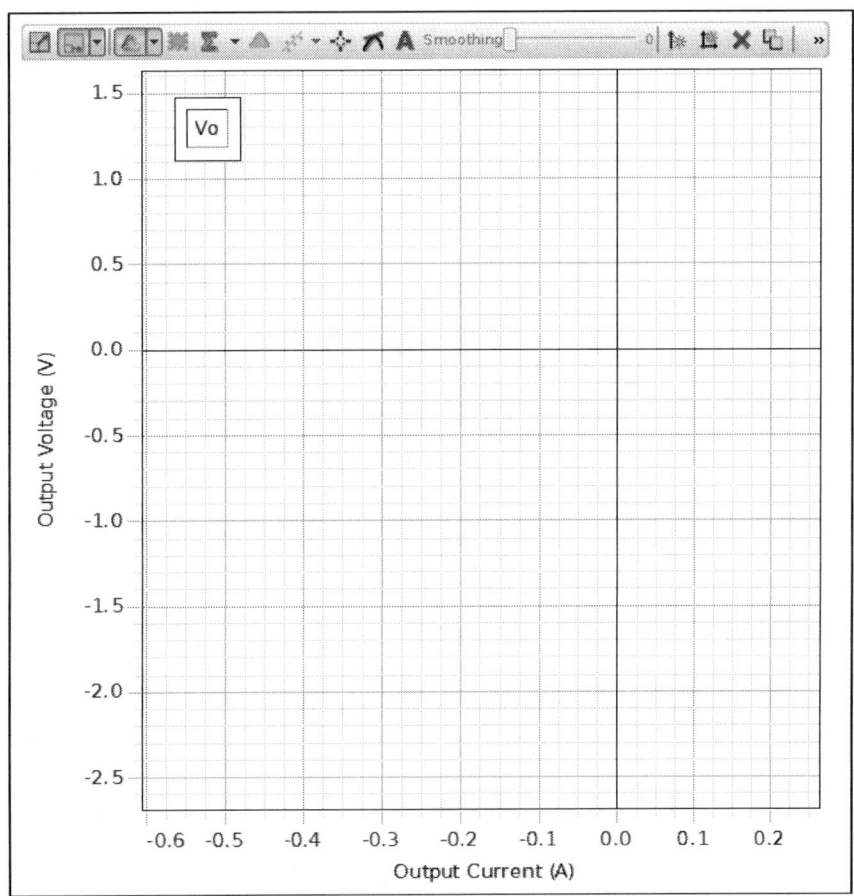

Graph 2: Output Voltage vs. Output Current for the Light Bulb

Analysis: Diode

1. Click on the black triangle by the Run Select icon and select the "diode up", and "diode down" runs. Both curves should show at the same time. Sketch the data in Graph 3.
2. Click on the Scale to **Fit** icon.
3. Select the data between +/- 1 V and +/- 0.05 A with the Selection box
4. Click the Scale to **Fit** icon.
5. Click the Remove Active Element icon.
6. Click the black triangle by the Run Selection icon and select "330 Run" in addition to the "diode up", and "diode down" runs. Note that all three curves have the same x-intercept (where V = 0). The current is assumed to be zero here, but may not read zero because of error in the current sensor.

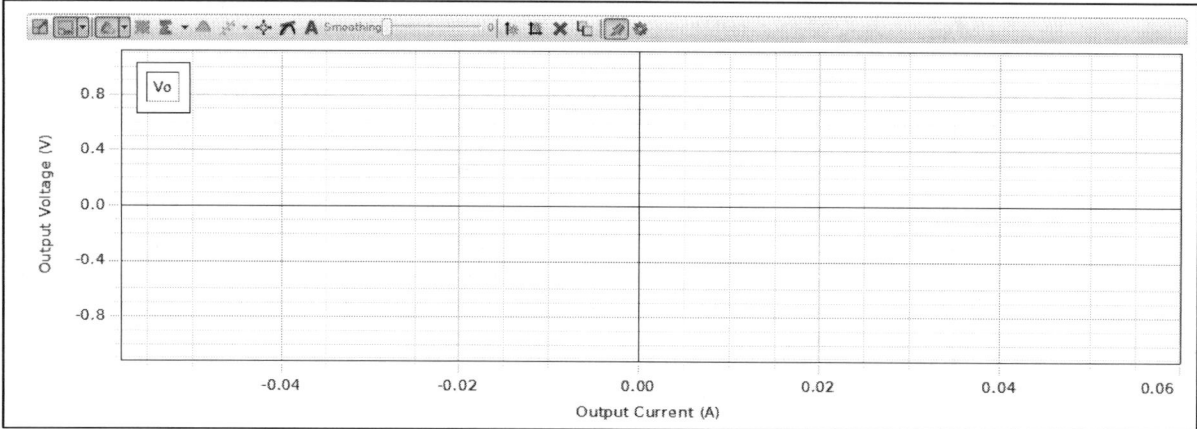

Graph 3: Output Voltage vs. Output Current for the diode

Last/First Name (print): _____

PHYS 1440-Section _____ Username ID (xxx9999): _____

Conclusions:

Verifying Ohm's Law for a Resistor

1. Why do the lines formed by the data have some width?

2. Why do the "330 Run" line and the "560 Run" line cross at the same place on the current axis?

3. How well do the resistors obey Ohm's Law? Explain fully how you know!

4. What is the physical meaning of the Linear Fit slopes in the resistor data? Hint: what are the units of the slope?

Current-Voltage Relationship for a Light Bulb

5. Is the resistance of the light bulb constant? Explain your answer.

6. Why is there a kink? Why are there two values of current for each voltage near the kink?

7. Are there any regions where the light bulb obeys Ohm's Law? Explain your answer.

Last/First Name (print): _____

PHYS 1440-Section _____   Username ID (xxx9999): _____

Current-Voltage Relationship for a Diode

8. Does the diode obey Ohm's Law? Explain your answer.

9. Does the diode conduct electricity for both positive and negative voltages in a single orientation? Explain your answer.

10. At what voltage does the diode conduct electricity?

1440 Questionnaire: Experiment 4: Ohm's Law

Do not put your name on this page. Hand in this page separately.

1. TA name(s):

2. What one thing did you like best about this laboratory?

3. What one thing did you like least about this laboratory?

4. What one thing would you change in this laboratory?

5. What one thing would you leave the same?

Additional **Comments?**

Last/First Name (print): _____

PHYS 1440-Section _____ Username ID (xxx9999): _____

Experiment 5:

Series and Parallel Circuits

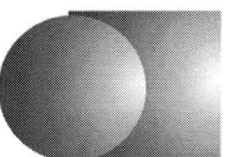

Pre-lab

1. What are two different ways that resistors can be connected together in a circuit? Explain the difference between the two in terms of calculating equivalent resistance.

2. Calculate the equivalent resistance of circuit diagram 1.

3. Calculate the equivalent resistance of circuit diagram 2. How much current would be drawn from a 12V battery connected to this circuit?

Last/First Name (print):

Experiment 5:
Series and Parallel Circuits

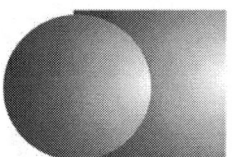

Introduction

The purpose of this experiment is to help the student understand series and parallel circuits, how to calculate their equivalent resistance, and how to construct them in the laboratory. The resistance of four circuits will be determined both theoretically and experimentally. The experimental resistance will be calculated by measuring both the voltage and current of the constructed circuits. The behavior of light bulbs connected in series and parallel will also be examined.

Theory

Calculating Equivalent Resistance

A resistor generally means a device that obeys Ohm's Law (many devices do not) and has a resistance R.

$$\text{Ohm's Law: } V = IR \qquad \textbf{Equation 1}$$

Two (or more) resistors can be connected in series (as in circuit diagram 1), or in parallel (as in circuit diagram 2). Resistors can also be connected in a series/parallel circuit as shown in circuit diagrams 3 & 4. An equivalent resistor is a single resistor that could replace a more complex circuit and produce the same total current when the same total voltage is applied. This is shown in Figures 1 and 2. For a series circuit, the resistances are additive:

$$R_{eq} = R_1 + R_2 \qquad \textbf{Equation 2}$$

where R_{eq} is the equivalent resistance.

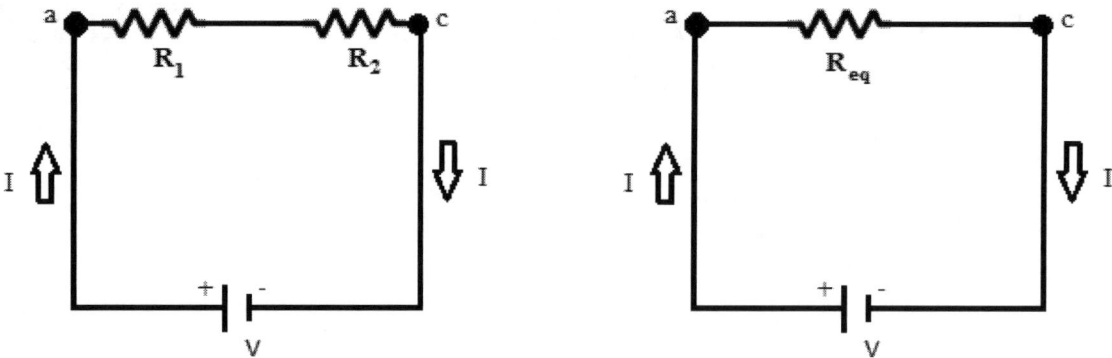

Figure 1: Resistors in Series

For a parallel circuit, the resistances add as reciprocals

$$\frac{1}{R_{eq}} = \frac{1}{R_1} + \frac{1}{R_2}$$ **Equation 3**

Remember, when adding fractions, they must have like denominators!

We must multiply each fraction so that they have common denominators.

$$\frac{1}{R_1} \cdot \frac{R_2}{R_2} + \frac{1}{R_2} \cdot \frac{R_1}{R_1} = \frac{R_2}{R_1 R_2} + \frac{R_1}{R_1 R_2} = \frac{R_1 + R_2}{R_1 R_2}$$

So we get that

$$\frac{1}{R_{eq}} = \frac{1}{R_1} + \frac{1}{R_2} = \frac{R_1 + R_2}{R_1 R_2}$$

If we take the reciprocal of both sides we obtain another expression for calculating equivalent resistance in parallel circuits.

$$R_{eq} = \frac{R_1 R_2}{R_1 + R_2}$$ **Equation 4**

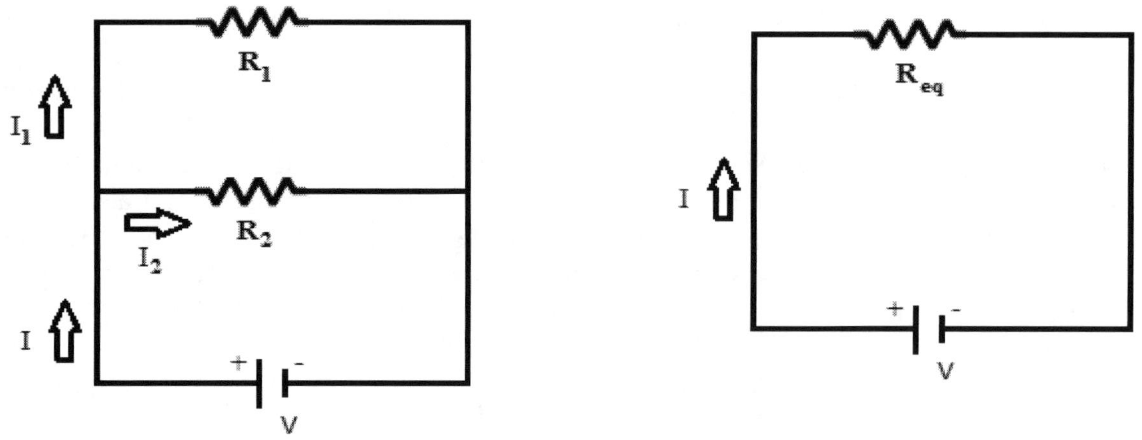

Figure 2: Resistors in Parallel

A more complex circuit like Circuit Diagram 3 can be handled by combining R_1 and R_2 into an equivalent resistance with Equation 4. That equivalent resistance is then put in series with R_3 and equation 2 is used to find the equivalent resistance for the whole circuit. Circuit Diagram 4 can be handled in a similar manner.

In series circuits the current is the same through each resistor, but the voltage drop across each resistor may be different. Likewise, in a parallel circuit the voltage drop across each resistor is the same, but the current through each resistor may be different.

Power
A simple understanding of power will help the student understand what is physically happening in this experiment. Power is the rate at which work is done for a system. Electrical power is defined as:

$$P = IV \qquad \textbf{Equation 5}$$

Where P is the power measured in watts, I is the current in amperes, and V is the voltage drop across the device measured in volts. It is useful to consider power in terms of current and resistance. Remember that Ohm's law relates voltage to current and resistance. If this is plugged into equation 4, another way of writing power is developed. This is only true for devices that obey Ohm's law!

$$P = IV = I(IR) = I^2 R \qquad \textbf{Equation 6}$$

R is the resistance measured in ohms. Power is directly proportional to resistance and the current squared. If two devices have the same resistance, but device 1 has twice as much current running through it compared to device 2, device 1 will have 4 times the power.

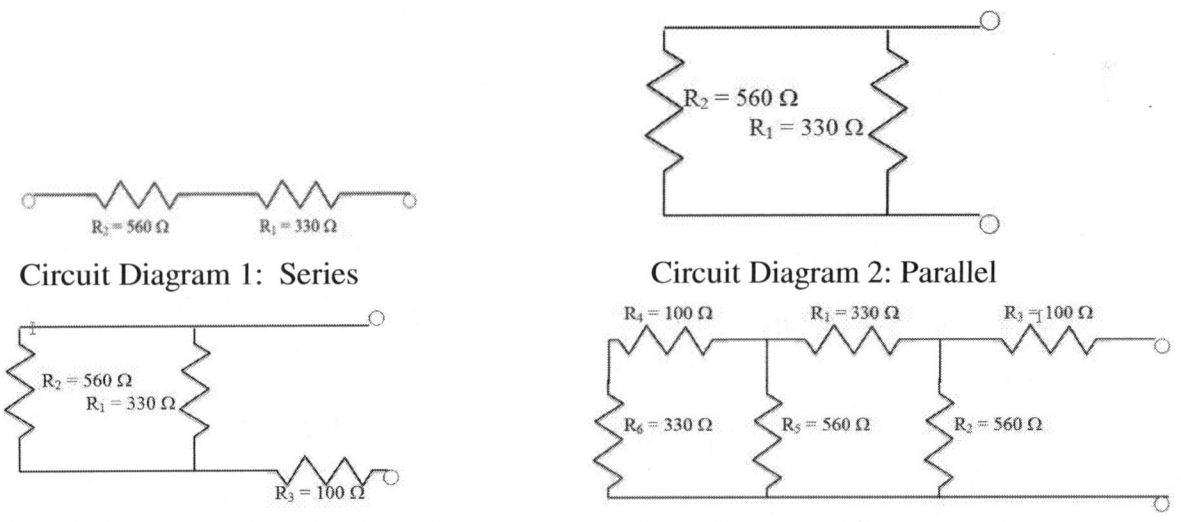

Circuit Diagram 1: Series

Circuit Diagram 2: Parallel

Circuit Diagram 3: Simple Series Parallel

Circuit Diagram 4: Complex Series Parallel

Equipment

1	AC/DC Electronics Laboratory	EM-8656
1	Short Patch Cords (set of 8)	SE-7123
1	850 Universal Interface	UI-5000
1	PASCO Capstone	UI-5400

Last/First Name (print): _____

PHYS 1440-Section _____ Username ID (xxx9999): _____

Procedure: Resistor Calibration

The gold band on each resistor conveys a precision of +/- 5%. By measuring each resistor using the Universal Interface, the precision can be improved to about +/- 1%. The 850 Universal Interface measures the voltage between the red and black patch cables and the current flowing through circuit.

1. Locate two 100 Ω (brown-black-brown-gold) resistors, two 330Ω (orange-orange-brown-gold) resistors, and two 560Ω (green-blue-brown-gold) resistors.
2. Label different positions on a piece of paper R_1, R_2, R_3, R_4, R_5, R_6.
3. Put the 100Ω resistors at R_3 and R_4, the 330Ω resistors at R_1 and R_6 places, and the 560Ω resistors at R_2 and R_5 places
4. Insert a resistor between the spring clips as shown in Figure 3.
5. Open the Signal Generator at the left of the page.
6. Set Waveform for a Triangle wave at 10 Hz and Amplitude for 10 V.
7. Click Auto.
8. Click Signal Generator to close the panel.
9. Click Record. Click Stop after 10 seconds.
10. Select the data and click Scale to Fit
11. Click black triangle by the Curve Fit icon on the graph toolbar and apply a Linear fit.
12. The slope of the line is the measured resistance. Enter the value in Table 1.
13. Label different positions on a piece of paper R_1, R_2, R_3, R_4, R_5, R_6.
14. Put the 100Ω resistors at R_3 and R_4, the 330Ω resistors at R_1 and R_6 places, and the 560Ω resistors at R_2 and R_5 places
15. Repeat for each of the six resistors. Make sure you keep track of which resistor you have measured.

Table 1: Measured Resistance Values

R_1 = 330 (Ω)	R_2 = 560 (Ω)	R_3 = 100 (Ω)	R_4 = 100 (Ω)	R_5 = 560 (Ω)	R_6 = 330 (Ω)

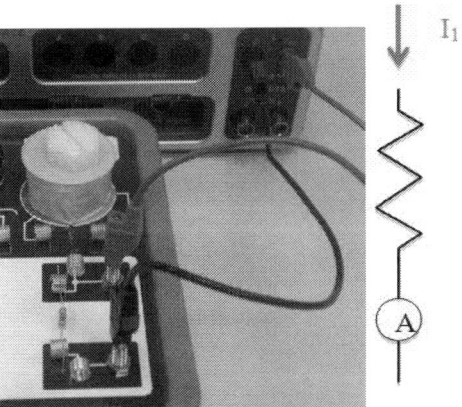

Figure 3: Calibration Circuit

Procedure: Ammeter Calibration

The internal ammeter works by measuring the voltage drop across a small resistor (~0.1 Ω). Since the sensitivity is about 0.1 mA, this means the 850 Universal Interfaces must measure voltages of 0.01 mV. Noise can result in significant zero error. By averaging over several seconds we can achieve a precision of 0.1-0.2 mA, but with systematic errors that can approach 5 milliamps. We can correct for this with a brief calibration procedure.

1. Insert the resistor with an actual resistance closest to 100 Ω into the calibration circuit (see Figure 3).
2. Correct the first column using Ohm's Law: $I = V/R$. Where V is the voltage of a particular run, and R is the resistance of the closest resistor to 100 Ω.
3. Click **open** the Signal Generator at the left of the screen.
4. Set Output 1 for a DC Waveform and a DC Voltage of 0 V.
5. Click the On button.
6. Click Record (bottom left of screen).
7. Wait several seconds until the measured current stops varying as the average becomes well defined.
8. Click Stop.
9. Enter the value in the second column of table 2.
10. Subtract the measured current (column 2) from the theory current (column 1) and enter the value into A Corrected (mA); column 3 = column 1 - column 2
11. Click Delete Last Run at the bottom of the screen.
12. In the Signal Generator panel, increase the voltage by 1 V and repeat. Then repeat steps 3-10, increasing the voltage by 1 V each time until 7 V is reached.

Table 2: Ammeter Calibration Data

Theory Current (mA)	A Current (mA)	A Corrected (mA)
=0V/R: 0 mA		
=1V/R:		
=2V/R:		
=3V/R:		
=4V/R:		
=5V/R:		
=6V/R:		
=7V/R:		

Note: If the calibration procedure is skipped, use the assumed values of the resistors. You may skip Tables 1 and 2. You may also ignore the "i Corrected" column in Table 3. Skip steps 9-11 in the series and parallel procedure. Skipping the calibration will induce error into the experiment. It is also important to note that the circuits with a higher equivalent resistance will produce a smaller current. In turn, a higher error will be present due to the current offset.

Setup: Series and Parallel Circuits

Construct the circuit shown in the Circuit Diagram 1 and Figure 4.

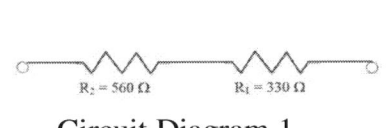

Circuit Diagram 1

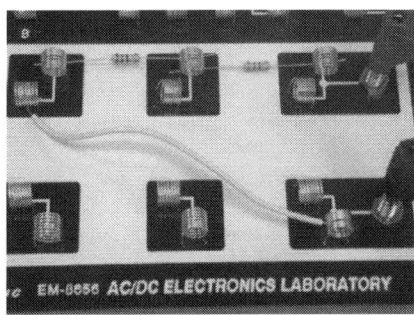

Figure 4: Series Circuit

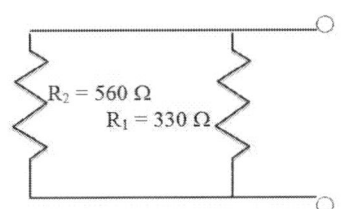

Circuit Diagram 2

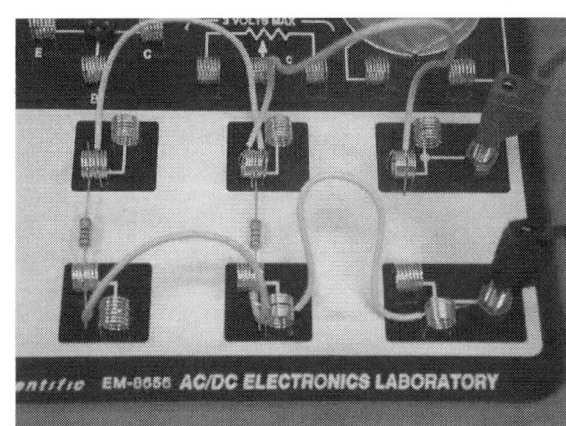

Figure 5: Parallel Circuit

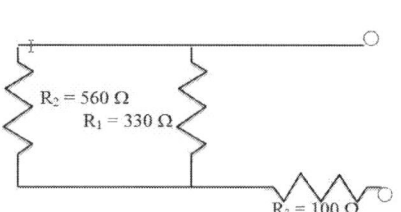
Circuit Diagram 3

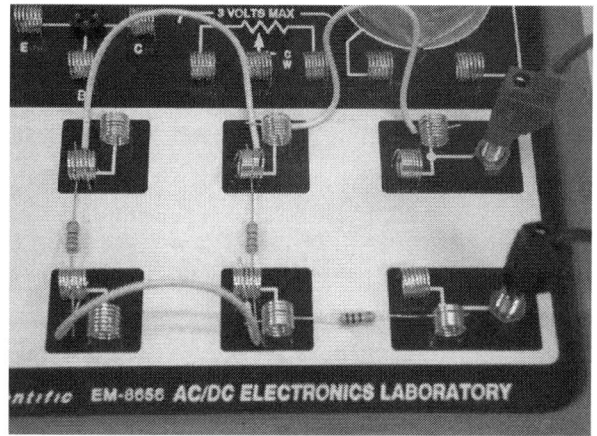
Figure 6: Simple Series Parallel

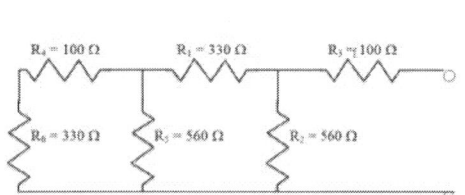
Circuit Diagram 4

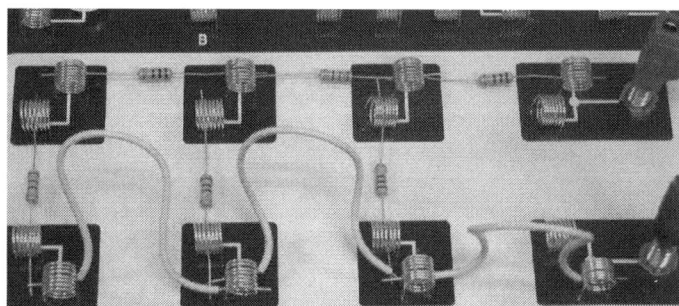
Figure 7: Complex Series Parallel

Last/First Name (print): _____

PHYS 1440-Section _____ Username ID (xxx9999): _____

Procedure: Series and Parallel Circuits

1. Open the Signal Generator.
2. Select Output 1.
3. Set DC Waveform at 15 V.
4. Click Auto.
5. Click the Signal Generator again to close the panel.
6. Click Record.
7. Click stop after the values stop varying.
8. Enter the value into i-Measured in Table 3.
9. Identify the closest current to i-Measured in the second column of Table 2..
10. Add the correction (Column 3 of Table 2) to i-Measured.
11. Enter this value in the i-Corrected column of Table 3.
12. Applying Ohm's Law: $R = \frac{V}{I}$; calculate the Measured Resistance using i-Corrected and the measured. Record in Table 3.
13. Repeat Steps 4-10 for circuit diagrams 2-4 which are shown constructed in Figures 5-7.
14. Calculate each theoretical resistance for circuit diagrams 1-4 and enter the value in the second column of Table 3.
15. Calculate the % Difference between theoretical Req and the Measured Req. Record this in Table 3.

Table 3: Resistance Summary

Circuit Diagram	Theoretical R_{eq} (Ω)	i Measured (mA)	i Corrected (mA)	Measured R_{eq} (Ω)	% Difference
1					
2					
3					
4					

Last/First Name (print): _____

PHYS 1440-Section _____ Username ID (xxx9999): _____

Setup: Light Bulb

1. Attach 2 long wires from the red and black leads to light bulb A as shown in Figure 8.
2. Open the Signal Generator at the left of the page.
3. Set Waveform for a Triangle wave at 10 Hz and Amplitude for 2.5 V.
4. Click Auto then close the signal generator.

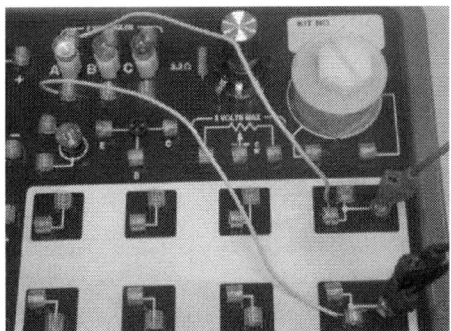
Figure 8: Single Light Bulb

Figure 9: Light Bulb A&B in Series

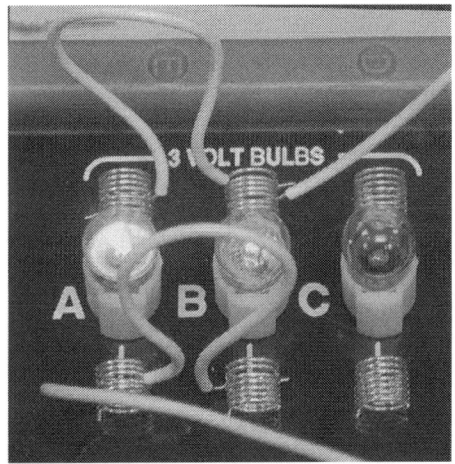
Figure 10: Light Bulb A&B in Parallel

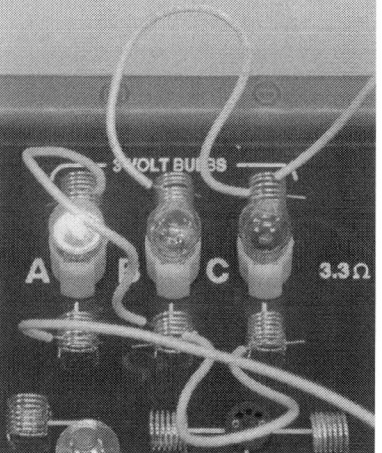
Figure 11: Series Parallel Light Bulb

Procedure: Light Bulb

1. Click Record.
2. Click Stop after a few seconds.
3. Select the data and apply a linear fit. The data will not be completely linear since the bulb resistance changes as it heats and cools.
4. Enter the value of the slope to the nearest 0.1 Ω for bulb in the Table 4. The slope of the line is the measured resistance of the bulb at its operating temperature.
5. Repeat for the other two bulbs in sockets B and C.
6. Put bulbs A&B in series as shown in Figure 9 on the previous page and repeat steps 1-4.
7. Put bulbs A&B in parallel as shown in Figure 10 on the previous page and repeat steps 1-4.
8. Put bulb A in series with bulbs B&C which are in parallel as shown in Figure 11 and repeat steps 1-4.
9. Sketch each data set in graph 4.

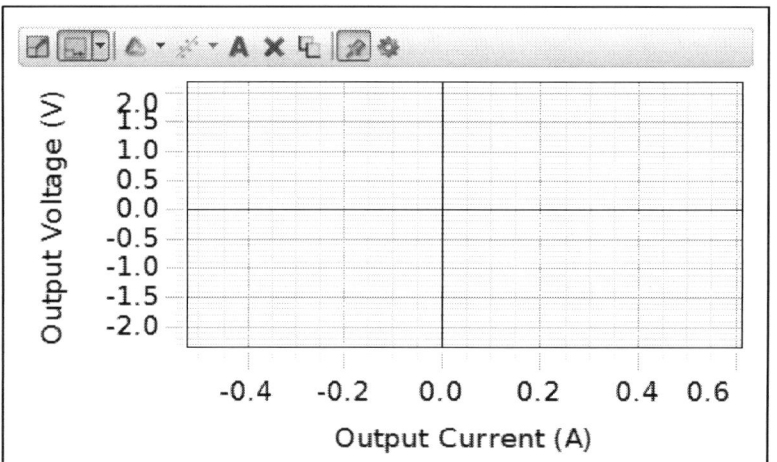

Graph 4: Output Voltage vs Output Current for Light Bulb Setups

Table 4: Bulb Resistance

Bulb System	Measured Resistance (Ω)
A	
B	
C	
A&B Series	
A&B Parallel	
Series Parallel	

Last/First Name (print): _____

PHYS 1440-Section _____   Username ID (xxx9999): _____

Conclusions

Series and Parallel Circuits

1. How well did the theoretical values compare to the experimental values?

2. What are sources of error in this experiment, and how could they be minimized?

Light Bulb

3. How does the brightness of two bulbs in series compare to a single bulb? Explain this in terms of power.

4. How does the brightness of two bulbs in parallel compare to a single bulb? Explain your answer.

5. Explain what is happening in the series parallel setup.

1440 Questionnaire: Experiment 5: Series Parallel Circuits

Do not put your name on this page. Hand in this page separately.

1. TA name(s):

2. What one thing did you like best about this laboratory?

3. What one thing did you like least about this laboratory?

4. What one thing would you change in this laboratory?

5. What one thing would you leave the same?

Additional Comments?

First Name (print): _____

PHYS 1440-Section _____ Username ID (xxx9999): _____

Experiment 6:
Capacitance and RC Circuit

Pre-lab, Part A: Capacitance

1. What is a capacitor? Define capacitance. How is the capacitance of a parallel plate capacitor related to the geometrical parameters (area and distance), and the space's relative permittivity.

2. Find the capacitance of a parallel plate capacitor that has two square plates both having a side length of 6 cm, where these plates are separated by 15 mm. Assume the insulating material between the two plates is a vacuum. Be sure to use the correct units in your calculations.

3. When can the approximation of infinite or large plates be applied?

Pre-lab, Part B: RC Circuit

4. What is the time constant of the RC circuit if R=1 kΩ and C=3μF.

5. Explain how to measure half-life of an RC circuit, using the decay of voltage across a capacitor.

Last/First Name (print): _____

Experiment 6:
Capacitance and RC Circuit

Introduction

The purpose of this experiment is to investigate how the capacitance of a parallel-plate capacitor varies when the plate separation is changed and to qualitatively see the effect of introducing a dielectric material between the plates. A computer model of the system will be developed to help the student better understand capacitance. The manner by which the voltage on a capacitor decreases is studied. Additionally, the half-life for the decay in a RC circuit is measured directly and also calculated using the capacitive time constant.

PART A: Capacitance

Theory

A capacitor is used to store charge and consequently energy. A capacitor can be made with any two conductive plates and a dielectric spacer between them. If the conductors are connected to a potential difference, V, such as the opposite terminals of a battery, then the two conductors are charged with equal but opposite amount of charge Q in the capacitor. The total charge on both plates of the capacitor is zero. The *capacitance* of the device is defined as the coefficient between an applied potential difference V and amount of charge Q stored in each plate:

$$Q = CV \qquad \textbf{Equation 1}$$

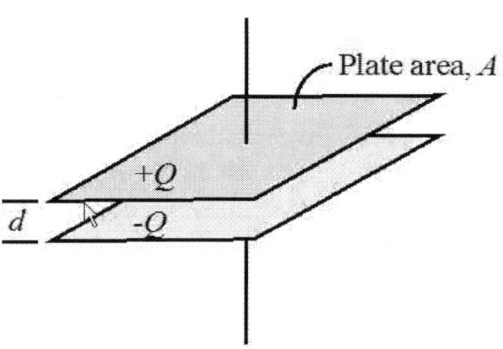

Figure 1: Sketch of a parallel-plate capacitor

The simplest form of a capacitor consists of two parallel conducting plates (Figure 1), each with area A, separated by a distance d. The charge is uniformly distributed on the surface of the plates. The capacitance of the parallel-plate capacitor is given by:

$$C = \frac{\kappa \varepsilon_0 A}{d} \qquad \textbf{Equation 2}$$

Where κ is the relative permittivity of the dielectric material between the plates ($\kappa = 1$ for a vacuum; $\kappa = 4.5$ for laminated paper; $\kappa = 2\text{-}6$ for wood depending on the density), and ε_0 is the permittivity of free space; $\varepsilon_0 = 8.85 \times 10^{-12}$ F/m. The SI unit of capacitance is the Farad (F).

Equation 2 is true in *the approximation of infinite (or large) plates*, where one can neglect *the edge effects,* formally meaning that $\sqrt{A} \gg d$.

Examination of Equations 1 and 2 shows that if the large plates approximation is valid, then V is directly proportional to d

$$V = \frac{Qd}{\kappa \varepsilon_0 A}$$ **Equation 3**

The system we use is more complex. In addition to the two moveable parallel plates, the connecting wires and the electrometer also have some capacitance. This capacitance is roughly equal to the capacitance of the moveable plates when the plates are 1 cm apart and cannot be ignored. Including this gives:

$$C = \kappa \varepsilon_0 A/d + C_{sys}$$ **Equation 4**

where C_{sys} is the capacitance of the rest of the system. Substitution of Equation 4 into Equation 1 yields:

$$V = Q/[\kappa \varepsilon_0 A/d + C_{sys}]$$ **Equation 5**

Any material placed between the plates of a capacitor will increase its capacitance by a factor κ called the dielectric constant where:

$$C = \kappa C_0$$ **Equation 6**

with C_0 being the capacitance when there is a vacuum between the plates of the capacitor.

Dielectric materials are non-conductive. Any dielectric material can be used to keep the plates in a capacitor insulated from each other (preventing them from touching and discharging). To three significant figures, $\kappa = 1.00$ for air. For all materials, $\kappa > 1$. If the charge on a capacitor is kept constant while a dielectric is inserted between the plates, Equations 1 & 6 yield:

$$Q = CV = C_0V_0 = (C/\kappa)V_0, \text{ so } V = V_0/\kappa$$

Where V_0 is the voltage before inserting the dielectric and V is the voltage after insertion. Since κ is always greater than 1, we have

$$V < V_0$$ **Equation 7**

Equipment

1	Basic Electrometer	ES – 9078
1	Basic Variable Capacitor	ES – 9079
1	Electrostatics Voltage Source	ES – 9077
1	Short Patch Cords (set of 8)	SE-7123
1	Resistor/Capacitor/Inductor Network	UI-5210
1	Voltage Sensor	UI-5100
1	Short Patch Cord Set	SE-7123
1	850 Universal Interface	UI-5000
1	PASCO Capstone	UI-5400

Setup A: Capacitance

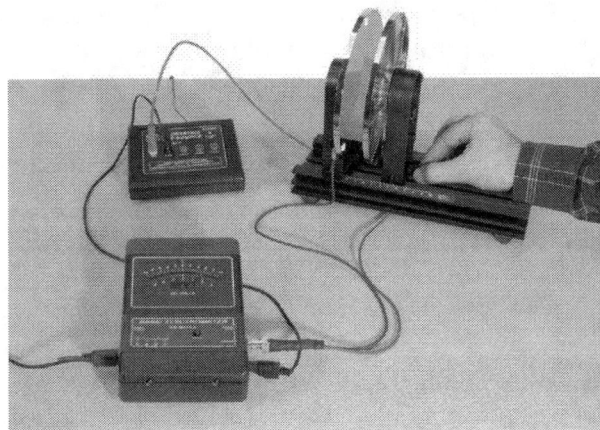

Figure A1: Setup

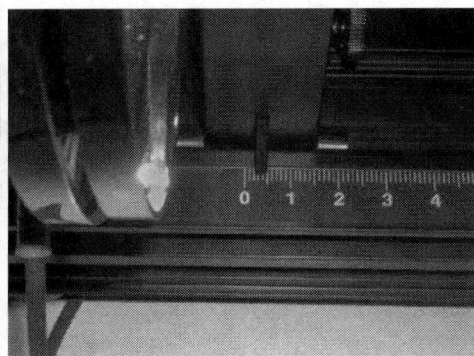

Figure A2: Indicator Foot

1. Move the Variable Capacitor plates so they are about 2 mm apart.
2. Adjust screws on the back of the moveable plate to make the plates parallel.
3. Position the movable plate so the leading edge of the indicator foot (see Fig. A2) is at the 0.2 cm position. The gap between the two plates should be 2 mm all the way around.
4. Attach the twin lead (red & black) connector to the Signal Input jack on the Basic Electrometer.
5. Attach the black spade end of the twin lead to the fixed plate
6. Attach the red spade end of the twin lead to the movable plate
7. Attach a black banana/banana wire as shown from the common (com) terminal on the Electrostatic Voltage Source to the ground terminal on the Electrometer.
8. Attach the red banana/banana lead to the +30V terminal and leave one end free.
9. Attach an adaptor cable from the Electrometer signal output to the universal interface analog input A.
10. Plug in the power supply for the Electrostatic Voltage Source.
11. Shift the switch on the back to the On position. The green Power On light should glow.

Important Note:

- The edge effects are significant if the approximation of large plates relative to the gap size is not fulfilled. If the gap increases the dependence of potential difference of initially charged capacitor versus the gap size bend from linear dependence.
- If the edge effects are significant the measured potential difference will be sensitive to environment near the capacitor, for example: how the wires are routed and how far away from where your hand and your body will be. Note, people are conducting plates and have a significant amount of capacitance.
- The charges in this experiment all small so static discharge will result in the decreasing of measured voltage in time

Basic Variable Capacitor

The PASCO experimental Variable Capacitor consists of two metal plates 17.7 cm (7 in) in diameter with a plate area $A = 2.46 \times 10^{-2}$ m^2.

Last/First Name (print): ▒▒▒▒▒▒▒▒▒▒▒▒▒▒▒▒▒▒▒▒▒▒▒▒

PHYS 1440-Section ▒▒▒▒▒▒▒   Username ID (xxx9999): ▒▒▒▒▒▒▒▒▒▒▒▒

Procedure A1: **The Effect of the Plate Separation**

1. Set the capacitor plates 0.3 cm apart by setting the movable plate so leading edge of its indicator foot is at the 0.3 cm mark.
2. Turn on the electrometer and set the range button to the 100 V scale.
3. Remove any charge from the capacitor by momentarily touching both plates at the same time with a wire.
4. Zero the electrometer by pressing the 'ZERO' button until the needle goes to zero.
5. Momentarily connect a cable from the +30-V outlet in the voltage source to the stud on the back of the movable capacitor plate. This will charge the capacitor.
6. Remove the charging cable.
7. **Read all of the following steps.** They need to be performed quickly since the charge will slowly escape from the electrometer, especially if the humidity is high. One person should run the computer while one moves the capacitor plate. Everyone else should stay back. Everyone should try to be in the same position for each reading. Anybody too close can make the readings change.
8. Open the Data Tab, but read the rest of this page first.
9. Slide the movable plate so it is at 8.0 cm (leading edge of the indicator foot). Once the plate is in position, the person moving the plate should move away 50 cm or so and try to be in the same position for each measurement.
10. Click the PREVIEW button at the lower left to begin collecting data. Colored numbers will appear in first row of the table. The person doing the computer should click the Keep Sample (red checkmark in the lower left) button. The number in the first row will turn black and the colored number will move to the second row. The person at the computer should read the next separation (7 cm) out loud and wait.
11. Move the plate to 7.0 cm and repeat the process until 0.3 cm.
12. Click the STOP button to end the data collection.
13. Examine the graph. If the data looks like a smooth curve, continue to analysis. If not, repeat the process until the data is a smooth curve.
14. Record the data in Table 1.

Table 1: Air Gap Capacitor Data

Separation (cm)	Voltage (V)
8.0	
7.0	
6.0	
5.0	
4.0	
3.0	
2.0	
1.5	
1.0	
0.5	
0.3	

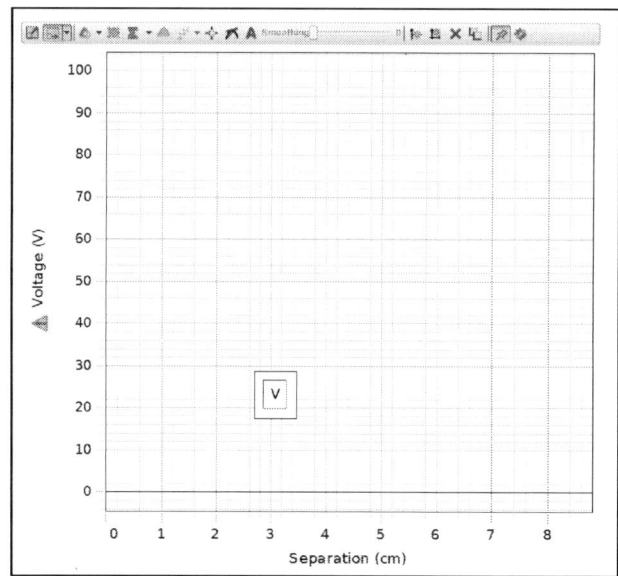

Figure A3: Air Gap Capacitor

Analysis A1: The Effect of the Plate Separation

This analysis will use equations 4 and 5 to show how our capacitor responds for a distance larger than the square of the plate area. The data obtained in procedure A1 is clearly not linear. If the large plate approximation was valid for this system, (Equation 3) V would be directly proportional to distance. The Voltage *vs.* Separation graph on the data page would be a straight line. When the gap distance became greater than a certain size the large plate approximation fails and Equation 4 must be used. To verify Equation 4 for the case where C is not zero, Q and C_{sys} must be known. This will be determined by fitting a math model (Equation 4) to the data. The calculator in the software will be used to fit a model to the data obtained in procedure A1. There are 4 lines in this calculator.

Line 1 is the charge of the system in Coulombs. Line 2 contains the physical parameters for this particular plate capacitor. Line 3 is the capacitance of the system in Farads. Line 4 is the voltage output from the model (Equation 5). Initial values for both line 1, 2, and 3 are already entered. The initial values for lines 1-3 have been calculated assuming an ideal 30 volts, no charge loss, and ideal capacitance for the system. Each system will be different due to charge loss, and slight deviations in capacitance and voltage. Lines 1 and 3 need to be modified as to fit the software model to the obtained data. Do not modify Line 2 as it is the calculated constants in this system!

1. Click open the Curve Fit Page
2. Use the Run Select button on the graph toolbar to select the best run
3. Modify the coefficient before the power of both lines 1 and 3 until the software generated model fits the data from procedure A1.
4. Record the Values for lines 1-3 in table 2.
5. Observe the potential decrease in time due to static discharge, if all the settings are the same.
6. Observe the edge effects when the V *vs.* d is not linear with the gap size.

Table 2: Software model

Line #	Value	Units
Line 1		Coulomb
Line 2		F Cm
Line 3		Farad

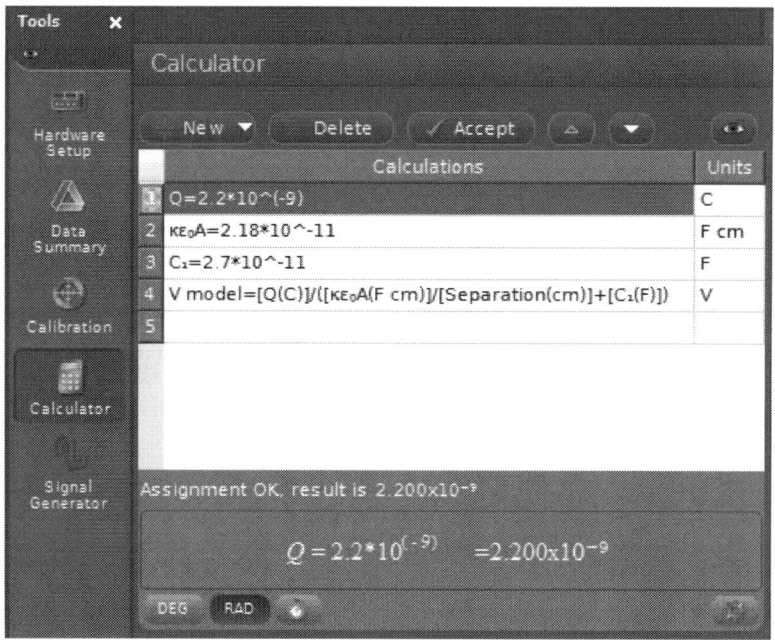

Figure A4: Calculator Setup

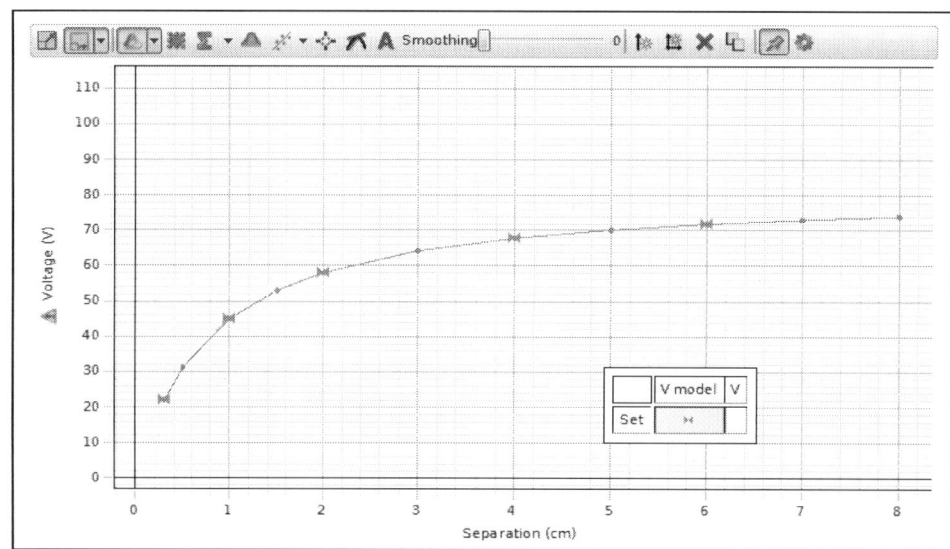

Figure A5: Capacitor Fit Model

Last/First Name (print): ▬▬▬▬▬▬▬▬▬▬▬▬▬▬▬▬▬▬▬▬▬▬▬

PHYS 2240-Section ▬▬▬▬▬ Username ID (xxx9999): ▬▬▬▬▬

Procedure A2: **The Effect of a Dielectric between the Plates**

This procedure will verify equations 6 and 7

1. You will use paper as the dielectric to be inserted between the plates. Use 0.5 cm of paper in a lab manual.
2. Position the movable plate of the capacitor at 0.5 cm.
3. Turn on the electrometer and set the range button to the 100 V scale.
4. Remove any charge from the capacitor by momentarily touching both plates at the same time with a wire. It is also best practice to keep hands away from the capacitor.
5. Zero the electrometer by pressing the 'ZERO' button. The needle must be at zero.
6. Momentarily connect a cable from the +30 V outlet in the voltage source to the stud on the back of the movable capacitor plate. This will charge the capacitor. Remove the charging cable.
7. Move the movable plate to 8 cm.
8. Click on the PREVIEW button below.
9. One student holds the paper directly above the gap between the capacitor plates so that the long side of the paper is vertical.
10. Hold the paper with one hand and keep the other hand on the metal connector attached to the signal input of the Electrometer so that there is no static charge on the student holding the paper.
11. Press the Keep Sample button to record the voltage when the paper is not between the plates.
12. Lower the paper between the two plates until it touches the base. Do not let the paper touch either plate! Keep your hand as far above the plates as possible.
13. Press the Keep Sample button to record the voltage when the paper is between the plates.
14. Pull the paper back above the plates and repeat steps 4 more times.
15. Click the STOP button to stop monitoring the data.
16. If the final voltage with the paper out is much different from the initial paper out value, you probably touched the plates and should repeat the experiment.
17. Record the data in Table 3.

Table 3: Dielectric Data

Trial	Paper Position	Voltage (V)
1	Out	
2	In	
3	Out	
4	In	
5	Out	
6	in	
7	Out	
8	in	
9	Out	

PART B: RC Circuit

Theory

In the RC circuit the voltage provided by the voltage source is equal to the potential drop across the capacitor plus the potential drop across the resistor (Kirchhoff's Voltage Law):

$$V_0 = V_C + V_R \qquad \textbf{Equation 8}$$

Therefore,

$$V_0 = Q/C + IR \qquad \textbf{Equation 9}$$

where C is the capacitance in Farads, Q is the charge in Coulombs, and V is the voltage in Volts. This equation contains the charge Q and the current, I, which, by definition, is the instantaneous rate of change in the charge. $I = \frac{dQ}{dt}$ (in mathematics it is called "first derivative of Q with respect to t").

Implying that the source voltage V_0 is off at some instant (back side of the square pulse), the voltage across the capacitor, $V_C(t)$ will decay exponentially with the characteristic time $\tau=RC$. RC has the units of seconds. The goal of this experiment is to obtain a better understanding of how a capacitor discharges with respect to its time constant, RC, and how RC is related to the half life.

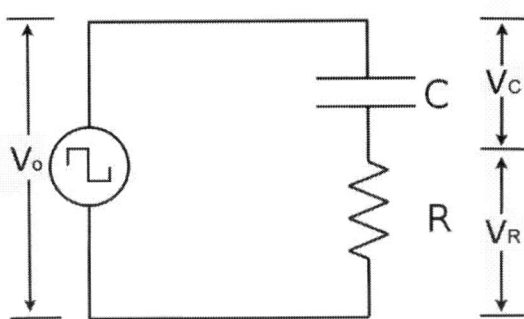

Figure B1: RC Voltages

Further analysis yields an expression describing the capacitor's voltage decay with respect to time:

$$V_C(t) = V_0 e^{-(t/RC)} \qquad \textbf{Equation 10}$$

where $V_0 = \dfrac{Q_{max}}{C}$. The rate that voltage across a capacitor (and the charge stored in the capacitor) decreases, depends on the resistance and capacitance of the circuit. If a capacitor is charged to an initial voltage, V_0, and is allowed to discharge through a resistor, R, the voltage, $V_C(t)$, across the capacitor will decrease exponentially.

The half-life, $t_{1/2}$ is defined to be the time that it takes for the voltage to decrease by half:

$$V_{(t_{1/2})} = V_0/2 = V_0 e^{-(t_{1/2}/RC)} \qquad \textbf{Equation 11}$$

Solving for the half-life gives

$$t_{1/2} = RC \ln 2. \qquad \textbf{Equation 12}$$

Setup B: Half-Life of an RC Circuit

Time Constant of an RC Circuit
1. Construct the circuit shown in Figure 1. The voltage source is Signal Generator #1 on the 850 Universal Interface. C = 3900 pF and R = 47 kΩ.
2. Click on Signal Generator #1 to connect the internal Output Voltage-Current Sensor.
3. Set the signal generator to a 350 Hz square wave with 2.5 V amplitude and 2.5 V offset. This will make the square wave all positive with amplitude of 5 V.
4. Set the signal generator on Auto.
5. Plug the Voltage Sensor into Channel A.
6. Connect the voltage sensor across the capacitor.

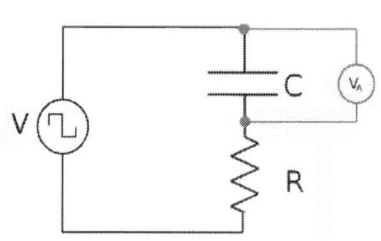
Figure B2. RC Circuit Diagram

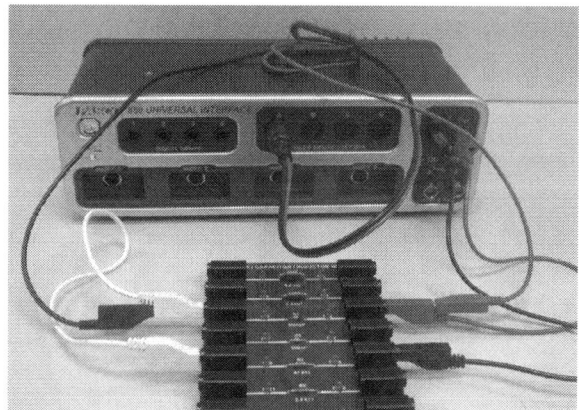
Figure B3. RC Circuit

Last/First Name (print): _____

PHYS 2240-Section _____ Username ID (xxx9999): _____

Procedure B1:

1. Click Monitor and adjust the scale on the oscilloscope so there is a complete cycle, so the capacitor fully charges and discharges.
2. Increase the number of points, using the tool on the scope toolbar, to the maximum allowed.
3. Take a snapshot (Create a Data Set on the graph toolbar) of both voltages shown. To take snapshots, select V_c (click on V on the legend) and click create Data Set.
4. Rename the snapshots "3900pF".

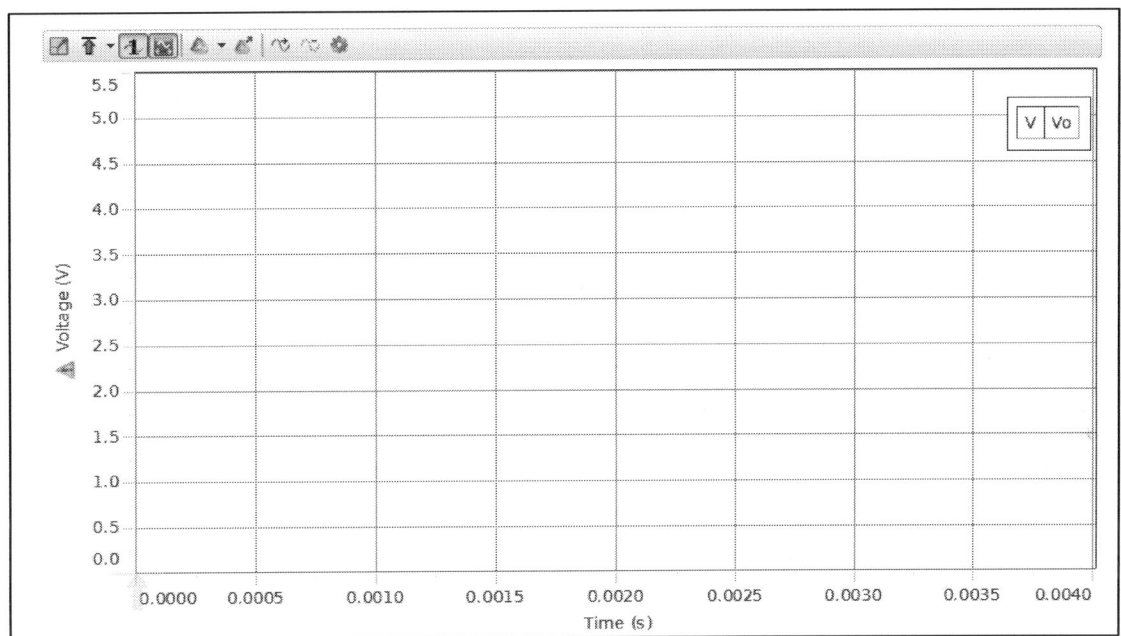

Figure B4: Scope showing the voltage across the capacitor and the output voltage

Procedure B2: Increase Voltage to 9 Volts

1. Change the output voltage to amplitude of 4.5 V with an offset of 4.5 V.
2. Keep the circuit the same.
3. Click Monitor and adjust the scale on the oscilloscope so there is a complete cycle, so the capacitor fully charges and discharges.
4. Increase the number of points, using the tool on the scope toolbar, to the maximum allowed. Then take a snapshot of both voltages shown. Rename the snapshot "9V".

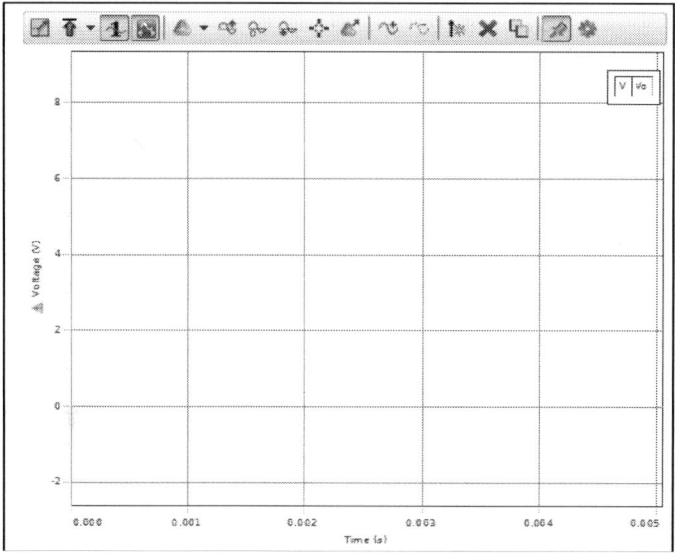

Figure B5: Scope showing output voltage and capacitor voltage for 3900pF, 9V

Last/First Name (print): _____

PHYS 2240-Section _____   Username ID (xxx9999): _____

Procedure B3: **Decrease Capacitance, 5 Volts**

Change C to 560 pF. Change square wave to 1800 Hz.
1. Change the initial sample rate to 500 kHz. Change the amplitude back to 2.5 V with an offset of 2.5 V.
2. Click Monitor and adjust the scale on the oscilloscope so there is a complete cycle, so the capacitor fully charges and discharges.
3. Increase the number of points, using the tool on the scope toolbar, to the maximum allowed.
4. Take a snapshot of both voltages shown. Rename the snapshots "560pF".

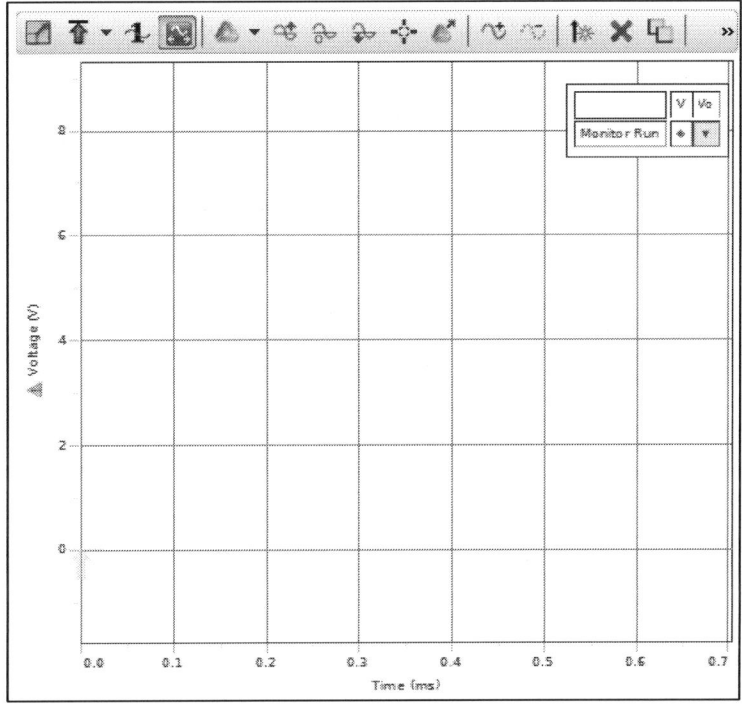

Figure B6: Scope showing output voltage and capacitor voltage for 560pF, 5V

Analysis B1: **3900 pF Capacitor, 5V**

1. Select the voltages for the 3900 pF run on the graph in Figure B7.
2. Adjust the horizontal scale so that the region from 4 V to 0.5 V on the capacitor voltage (V) nearly fills the field (see Figure B6).
3. Using the Coordinates Tool, measure the time it takes for the voltage to decay from 4V to 2V. This time is the half-life.
4. Reduce the snap-to-pixel distance to 1 in the properties of the Coordinates Tool (right click on the tool). It is convenient to arrange the coordinate tool as in Figure B6.
5. Measure the time it takes for the voltage to decay from 2V to 1V. This is still the half-life.
6. Measure the time it takes for the voltage to decay from 1V to 0.5 V. This is still the half-life.
7. Calculate the theoretical half-life.
8. Take the average of the three measured values and compare to the theoretical value using a percent difference.

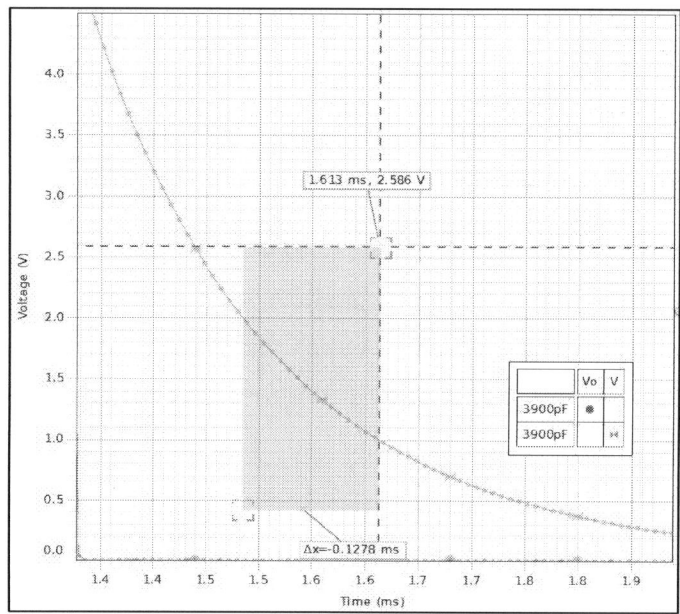

Figure B7: Graph and Coordinate Tool

Half Life	(ms)
$t_{1/2}$ (4V-2V)	
$t_{1/2}$ (2V-1V)	
$t_{1/2}$ (1V-0.5V)	
Average $t_{1/2}$	
$t_{1/2}$: Equation 12	
% difference:	

s

Last/First Name (print): _____

PHYS 2240-Section _____ Username ID (xxx9999): _____

Analysis B2: Linear Analysis 3900pF Capacitor at 5V

1. Choose the 3900pF run for the $\ln(V/V_0)$ and the voltage across the capacitor in the graph.
2. Select (using the Selection tool) the part of the $\ln(V/V_0)$ plot where the capacitor is discharging from 4 V to 0.5 V and fit it to a straight line using the Curve Fit tool. We avoid the region above 4 V since the square wave does not actually switch instantaneously. We limit to voltages above 0.5 V because at low voltage, system noise becomes important. This shows clearly in the divergence of the data from the straight line as the voltage goes to zero.
3. Use the slope of the line to find the half-life. To determine how the slope is related to the half-life, solve Equation (10) for $\ln(V/V_0)$.
4. Does this value agree with the value found from the previous analysis?

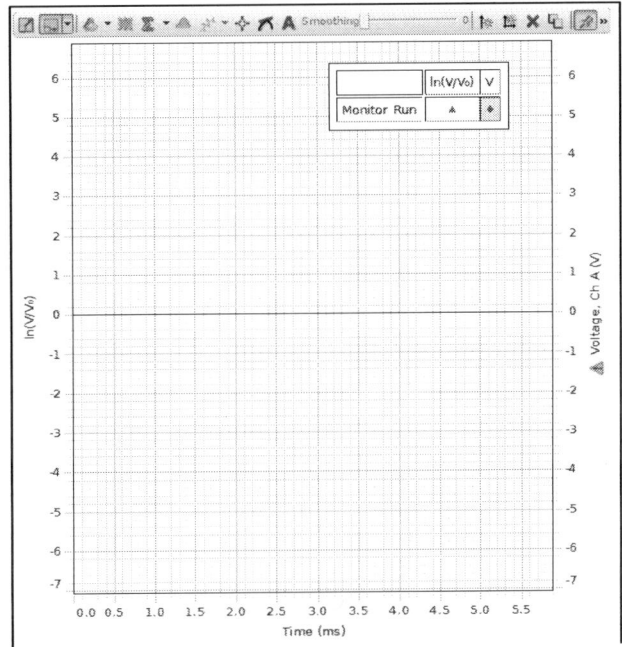

Figure B8: Linearized Data

1. $\ln(V/V_0) = -(1/RC)t + b =$

2. $t_{1/2} = RC\ln 2 =$

Analysis B3: 3900 pF Capacitor at 8V

1. Select the voltages for the 8V run on the graph in Figure B8. Adjust the horizontal scale so that the region from 8 V to 1 V on the capacitor voltage (V) nearly fills the field
2. Using the Coordinates Tool, measure the time it takes for the voltage to decay from 8 V to 4 V. This time is the half-life.
3. Measure the time it takes for the voltage to decay from 4 V to 2 V.
4. Measure the time it takes for the voltage to decay to 2 V to 1 V.
5. Has the theoretical half-life given by Equation (12) changed? Take the average of the three measured values of the half-life and compare to the theoretical value using a percent difference.

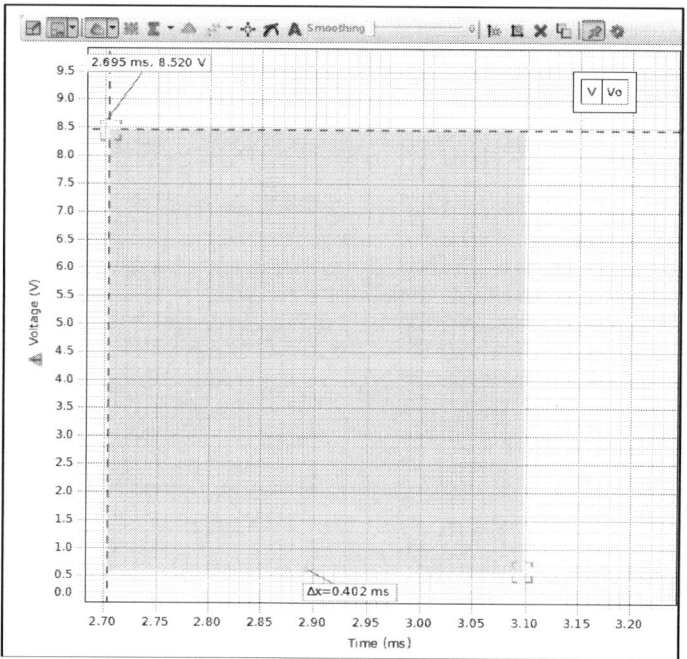

Figure B9: Analysis of Higher Voltage

Half Life	(ms)
$t_{1/2}$ (8V-4V)	
$t_{1/2}$ (4V-2V)	
$t_{1/2}$ (2V-1V)	
Average $t_{1/2}$	
$t_{1/2}$: Equation 12	
% difference:	

Last/First Name (print): _____

PHYS 2240-Section _____ Username ID (xxx9999): _____

Analysis B4: 560 pF Capacitor at 4V

1. Select the voltages for the 560pF run on the graph in Figure B10.
2. Measure the time it takes for the voltage to decay 4 V to 2 V by using the Coordinate Tool. This time is the half-life.
3. Measure the time it takes for the voltage to decay from 2 V to 1 V
4. Measure the time it takes for the voltage to decay from 1 V to 0.5 V.
5. Take the average of the three measured values of the RC and compare to the theoretical value using a percent difference.

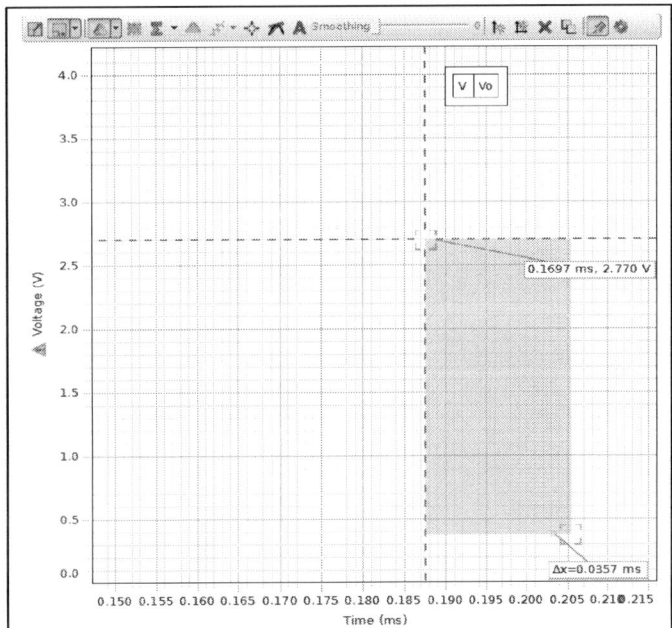

Figure B10. Analysis of Lower Capacitance

Half Life	(ms)
$t_{1/2}$ (4V-2V)	
$t_{1/2}$ (2V-1V)	
$t_{1/2}$ (1V-0.5V)	
Average $t_{1/2}$	
$t_{1/2}$ Equation 12	
% difference:	

Last/First Name (print): ▮

PHYS 2240-Section ▮   Username ID (xxx9999): ▮

Conclusions

Capacitance

1. What happened to the voltage *vs.* gap dependence as the plates got far from each other (d increasing)? Why did this happen?

2. Basing the conclusion on the data in Table 3, what does a dielectric do?

RC Circuit

3. Summarize how changing the capacitance changes the half-life. How does changing the voltage change the half-life?

4. Include the values found for the half-life and the % differences. Are the percent differences between theory and experimental data reasonable? Explain what causes the differences, and any sources of error.

1440 Questionnaire: Experiment 6: Capacitance and RC Circuit

Do not put your name on this page. Hand in this page separately.

1. TA name(s):

2. What one thing did you like best about this laboratory?

3. What one thing did you like least about this laboratory?

4. What one thing would you change in this laboratory?

5. What one thing would you leave the same?

Additional Comments?

Last/First Name (print): _____

PHYS 1440-Section _____ Username ID (xxx9999): _____

Experiment 7:
Kirchhoff's Circuit Laws

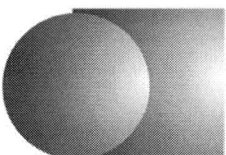

Pre-lab

1. What is Kirchhoff's Junction Law?

2. What is Kirchhoff's Loop Law?

3. Using Figure 1, apply both of Kirchhoff's laws and find I_3, ε_1, and ε_3. Let $I_1 = 3A$, $I_2 = 2A$, and $\varepsilon_2 = 10$ V.

Last/First Name (print): _____

Experiment 7:
Kirchhoff's Circuit Laws

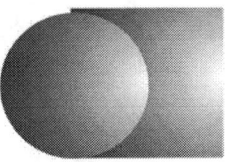

Introduction

This experiment will help the student understand Kirchhoff's circuit laws and how to solve electrical circuits using them. Both DC and AC circuits will be analyzed in context of Kirchhoff's current and voltage laws. A DC "Y" circuit will be analyzed, then an AC RC circuit. Kirchhoff's laws will be verified for both cases.

Theory

Kirchhoff's laws state:

1. *Junction Law: the total current flowing into and out of any junction point is zero at all times.*

$$\sum_i I_i = 0 \qquad \textbf{Equation 1}$$

2. *Loop Law: the sum of the voltage drops across any circuit element when one travels around any closed loop must be equal to zero.* Note that the voltage drop is negative if the voltage decreases and positive if the voltage increases in the travel direction that one goes around the loop.

$$\sum_i V_i = 0 \qquad \textbf{Equation 2}$$

Example Circuit

The following example will help the student realize how to use Kirchhoff's laws. The exemplary circuit shown in Figure 1 has two closed loops. To solve the system three simple steps are made:

1. Select the travel direction in each loop (shown by the red arrows; selection is arbitrary and does not affect the result.
2. Select the current directions in all braches if they are not given (in the case of Figure 1 circuit we selected direction of I_3 to be pointed down; note, the selection is arbitrary again: if the solution would give a negative answer for the unknown I_3, you should conclude that the physical direction is opposite to the initially selected).
3. Write the equations for each closed loop and junction using Equation 1 and 2.

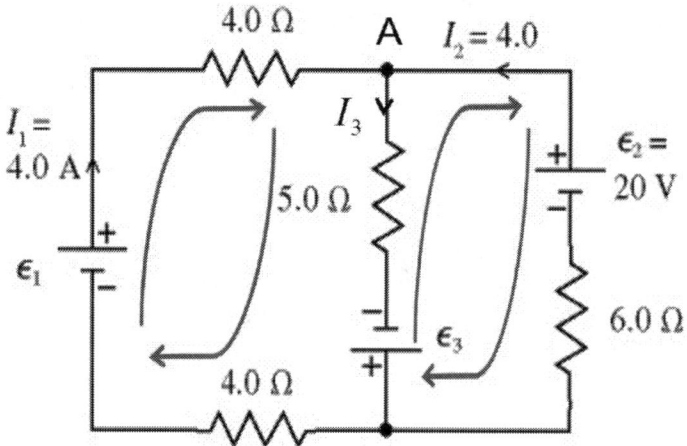

Figure 1. Exemplary two-loop circuit.
Note: The unknown electromotive forces (batteries voltage) and currents will be found.

Step 1 and step 2 are done and shown in Fig. 1. Let us move on to step 3 and write Kirchhoff's laws for this system.

System of Equations for Figure 1
  Junction A: $\quad I_1 + I_2 - I_3 = 0; \quad \rightarrow \quad I_1 + I_2 = I_3$
  Left Loop: $\quad -\epsilon_1 + I_1 R_{4\Omega} + I_3 R_{5\Omega} - \epsilon_3 + I_1 R_{4\Omega} = 0; \quad \rightarrow \quad 2 \cdot I_1 R_\Omega + I_3 R_{5\Omega} - \epsilon_1 - \epsilon_3$
  Right Loop: $\quad \epsilon_2 - I_2 R_{6\Omega} + \epsilon_3 - I_3 R_{5\Omega};$

Solution: Enter the values of the known variables and solve the system.

Junction A: $\quad 4A + 4A = I_3; \rightarrow I_3 = 8A$
Left Loop: $\quad 2 \cdot 4A \cdot 4\Omega + 8A \cdot 5\Omega - \epsilon_1 - \epsilon_3 = 0; \rightarrow 32V + 40V - \epsilon_1 - \epsilon_3 = 0$
Right Loop: $\quad 20V - 4A \cdot 6\Omega + \epsilon_3 - 8A \cdot 5\Omega = 0; \rightarrow 20V - 24V + \epsilon_3 - 40V = 0;$

$$\epsilon_3 = 40V + 24V - 20V = 44V$$

Plugging the value back into ϵ_3 back into the left loop yields the value of ϵ_1.

$$32V + 40V - \epsilon_1 - 44V = 0; \rightarrow \epsilon_1 = 28V$$

For this experiment a "Y" circuit will be used to test and exam Kirchhoff's laws.

Kirchhoff's laws will be applied to the "Y" circuit shown in Figure 2. Vo will be the supply voltage from Output 1 of the 850 Universal Interface. Vc will be produced by the voltage divider setup shown in Figure 3. Vc will not be calculated, but will be measured directly. To solve for the three unknown currents, three equations are needed.

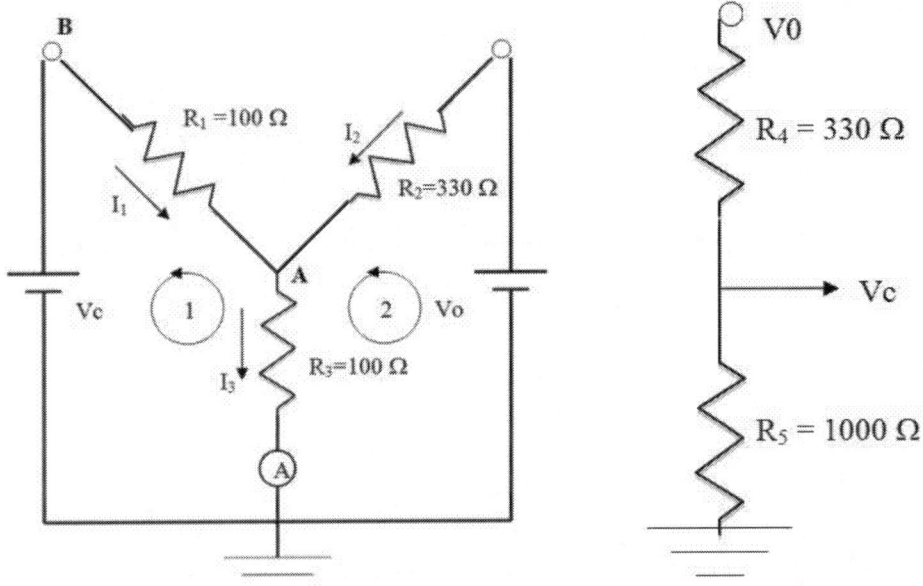

Figure 2: "Y" Circuit   Figure 3: Voltage Divider

Applying Equation 1 at node A in the circuit:

$$I_1 + I_2 - I_3 = 0 \qquad \textbf{Equation 3}$$

where current flowing in is considered positive and current flowing out is considered negative. Applying Equation 2 around loops 1 and 2 yields:

$$R_1 I_1 + R_3 I_3 - V_C = 0 \qquad \textbf{Equation 4}$$

$$-R_2 I_2 - R_3 I_3 + V_O = 0 \qquad \textbf{Equation 5}$$

where the voltage is negative if we go from high voltage to low (with the current arrow across a resistor).

The DC procedure will calculate I_1, I_2, and I_3. It is important to note that the experimental values for I_1 and I_2 will be calculated by measuring the voltage drop between V_C and V_A, and V_0 and V_A.

The experimental value for I_1:

$$I_1 = \frac{V_C - V_A}{R_1} \qquad \textbf{Equation 6}$$

The experimental value for I_2:

$$I_2 = \frac{V_0 - V_A}{R_2} \qquad \textbf{Equation 7}$$

Equipment

1	Resistive/Capacitive/Inductive Network	UI-5210
1	AC/DC Electronics Laboratory	EM-8656
3	Voltage Sensors	UI-5100
2	Current Probe	PS-2184
1	Short Patch Cords (set of 8)	SE-7123
1	850 Universal Interface	UI-5000
1	PASCO Capstone	UI-5400

Last/First Name (print): _____

PHYS 2240-Section _____   Username ID (xxx9999): _____

Setup: Calibration

Resistor Calibration

The color codes on the resistors only have a precision of +/- 5%. This can be improved to about +/- 1% using the calibration circuit.
1. Open the Signal Generator at the left of the page.
2. Set Waveform for 10 Hz and Amplitude for 10 V.
3. Click Auto.
4. Click Signal Generator to close the panel.
5. Click Record.
6. For each graph, click the black triangle by the Curve Fit icon on the graph toolbar and select Linear.
7. Average the values of the slope of each line (which is equal to the measured resistance) and enter them in the table.
8. Repeat for another 100 Ω and a 330 Ω (orange-orange-brown-gold) resistor. Make sure you keep track of which 100 Ω resistor is R_1 and which is R_3.

Table 1: Average Resistance Values

R1-100 (Ω)	R2-330 (Ω)	R3-100 (Ω)

Ammeter Calibration

Construct the circuit shown in Figure 4. Pick the resistor with the closest measured value to 100 Ω. The ≈ 100 Ω (+/- 5%) (brown-black-brown-gold) resistor is connected in series with a Current Probe (the ammeter, A with a circle around it, on the circuit diagram). A Voltage sensor is attached to the Current Probe as shown and then to the A Analog input on the 850 Universal Interface. It is important to observe polarity by connecting red to red and black to black where possible. There is a second ammeter built in to the 850 Output 1.

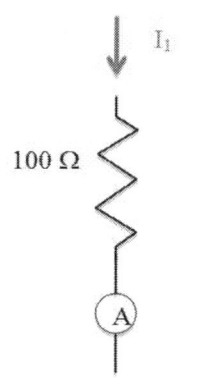
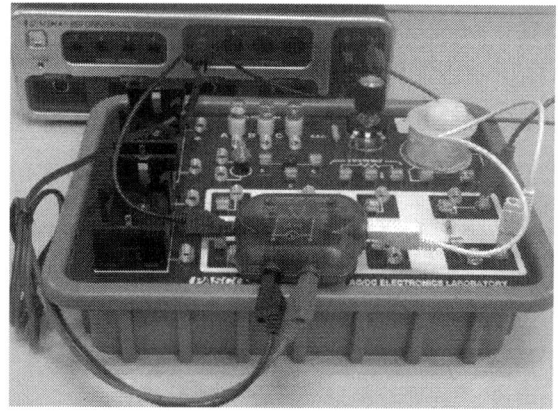
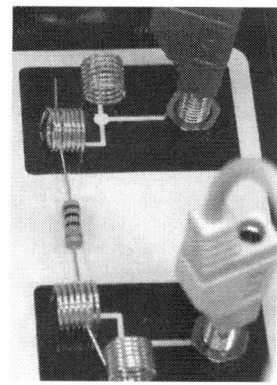

Figure 4: Calibration Circuit   Figure 5: Ammeter Calibration   Figure 6: 100 Ω Resistor

The internal ammeter and the external current probe work by measuring the voltage drop across a small resistor (~0.1 Ω). Since the sensitivity is about 0.1 mA, this means the 850 Universal Interfaces must measure voltages of 0.01 mV. Noise can result in significant zero error. By averaging over several seconds we can achieve a precision of 0.1-0.2 mA, but with systematic errors that can approach 5 milliamps. We can correct for this with a brief calibration procedure.

1. Insert the resistor with an actual resistance closest to 100 Ω into the calibration circuit (see Figure 4).
2. Correct the first column using Ohm's Law: $I = V/R$. Where V is the voltage of a particular run, and R is the resistance of the closest resistor to 100 Ω.
3. Click **open** the Signal Generator at the left of the screen.
4. Set Output 1 for a DC Waveform and a DC Voltage of 0 V.
5. Click the On button.
6. Click Record (bottom left of screen).
7. Wait several seconds until the measured current stops varying as the average becomes well defined.
8. Click Stop.
9. Enter the value in the second column of table 2.
10. Subtract the measured current (column 2) from the theory current (column 1) and enter the value into A Corrected (mA); column 3 = column 1 - column 2
11. Click Delete Last Run at the bottom of the screen.
12. In the Signal Generator panel, increase the voltage by 1 V and repeat. Then repeat steps 3-10, increasing the voltage by 1 V each time until 7 V is reached.

Last/First Name (print):

PHYS 2240-Section   Username ID (xxx9999):

Table 2: Ammeter Calibration Data

Theory Current (mA)	A Current (mA)	A Correct (mA)
=0V/R: 0 mA		
=1V/R:		
=2V/R:		
=3V/R:		
=4V/R:		
=5V/R:		
=6V/R:		
=7V/R:		

Setup: DC Current

1. Construct the circuit shown in Figures 2-3. The physical setup is shown in Figure 7. Note that the "Y" is on the left and the voltage divider (see Figure 6) is the two resistors on the right. R_1 is the 100 Ω resistor on the upper left. R_2 is the 330 Ω resistor at top center. R_3 is the 100 Ω resistor running from top to bottom. The voltage divider will supply V_c to point B on the left end of R_1.
2. Build this circuit using long wires.
3. Attach the ammeter by clipping the red end of the ammeter to the long wire attached to the bottom of R_3 as shown in Figure 8. Be careful with the alligator clips, as they could easily damage the springs!
4. Attach the black side of the ammeter to the wire coming from the lower left spring. This point will be used as ground.
5. Attach the black wire from Output 1 on the 850 to this point (ground).
6. Attach voltage Probes to Analog Inputs B & C on the Universal Interface.
7. Attach the black leads from the B & C inputs to the black side of Output 1 on the Universal Interface (ground)
8. Attach the red side of Output 1 to the upper banana input (V_o) on the circuit board as shown in Figure 9.
9. Attach the red lead from Analog Input C to the lower banana input (V_c) on the circuit board.
10. Attach the red lead from Analog Input B to the wire coming from the junction point A between the three resistors. It will measure V_a.

Figure 7: Circuit Setup

Figure 8: Hooking Up the Ammeter

Figure 9: Hooking Up the Voltage Probes

Last/First Name (print): _____

PHYS 2240-Section _____ Username ID (xxx9999): _____

If the calibration procedure is skipped, use the assumed values for the resistors. This will introduce an error in the experimental data. The I_3 corrected column in Table may be skipped along with step 10 in the DC Current Procedure. When the percent difference is calculated in the DC current analysis, use the uncorrected I_3 from table 3. Step 9 of the procedure below may be skipped.

Procedure: DC Current

1. Click open the Signal Generator at the left of the screen.
2. Set 850 Output 1 for a DC Waveform and a DC Voltage of 15 V.
3. Click the Auto button.
4. Click Record
5. Wait 5 seconds until the measured current stops varying as the average becomes well defined.
6. Click Stop.
7. Record the values for V_0, V_C, V_A, and I_3 in Table 3.
8. Calculate the experimental values for I_1 and I_2 using equations 6-7
9. Apply the Current Correction to I_3 and enter the value in I_3 corrected

Table 3: Experimental DC Circuit Voltages and Currents

V_o (V)	V_c (V)	V_a (V)	I_3 (mA)	I_3-Corrected (mA)	I_2 (mA)	I_1 (mA)

Analysis: DC Current

1. Using Equations 3-5 from Theory, calculate theoretical values for I_1, I_2, and I_3. These three equations may be solved algebraically, or by matrix methods using a 3x4 augmented matrix. Use The experimental V_0, V_C, and resistance values to solve for the current. Once solved, record in Table 4.
2. Calculate the % difference between the theoretical and experimental values and enter them in Table 4.

Table 4: Theoretical DC Circuit Currents and percent differences

I_3 Theoretical (mA)	I_2 Theoretical (mA)	I_1 Theoretical (mA)	%diff I_3 (%)	%diff I_2 (%)	%diff I_1 (%)

Last/First Name (print): _____

PHYS 2240-Section _____ Username ID (xxx9999): _____

Setup: RC with alternating current input

Construct the circuit shown in Circuit Diagram (Figure 10) with reference to Figures 11 & 12.

1. Construct the series circuit shown in Figure 11 using the 3900 pF capacitor and the 47 kΩ resistor. Note the polarities, with the red lead from the 850 Output 1 attached to the right side of the capacitor and the left side of the capacitor attached to the right side of the resistor.
2. Add the Voltage Sensors as shown in Figure 12. The polarities must match that shown in Circuit Diagram (Fig. 10) with the red leads on the left ends of the resistor and capacitor.
3. Attach the Voltage Sensor across the resistor to Analog Input C and the Voltage Sensor across the Capacitor must attach to Analog Input D.

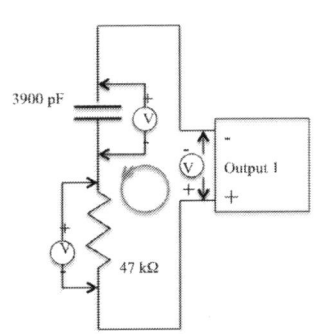      

Fig. 10: Circuit Diagram   Figure 11 RC Series   Figure 12: Adding the Sensors

Procedure: RC with alternating current input

1. Click open the Signal Generator. Make the signal a Square Waveform at 1000 Hz and 10 V. Click On.
2. Click Monitor at the lower left of the screen. The oscilloscope should record one cycle and stop. If any of the vertical jumps in the square wave fall exactly on one of the vertical time lines (0.0002 s, etc.), click monitor again.
3. Click Off on the Signal Generator and click on the Signal Generator button to close the Signal Generator panel.
4. The pattern on the oscilloscope shows the input voltage (V_0), the voltage across the resistor (V_R), and the voltage across the capacitor (V_c).

Analysis: RC with alternating current input

1. Click on the Coordinate tool on the RC Procedure page on the graph toolbar (crosshairs). Right click on the center of the cross-hairs on the graph and select Tool Properties and increase the Significant figures to 5.
2. Move the cross-hairs up the 0.2 ms (0.0002 s) line and record the values for V_0, V_R, & V_c to three decimal places in the RC Voltages table on this page. Try to get the value for time in the coordinates box as close to 2.00×10^{-4} s as possible (the last two digits will always be zero). You should get within 0.02×10^{-4} s always and generally exactly on 2.00×10^{-4} s. If you can't get it exactly on, try to take all three data points at the same time. You may need to move the Legend box in the upper right to see the data.
3. Repeat for 0.4 ms, 0.6 ms, 0.8 ms, and 1.0 ms.
4. Add V_0, V_R, and V_C. Enter this value in V loop.

Table 6: RC Voltages

Time (ms)	V_o (V)	V_r (V)	V_c (V)	V loop (V)
.2				
.4				
.6				
.8				
1.0				

Last/First Name (print): _____

PHYS 2240-Section _____ Username ID (xxx9999): _____

Conclusions

DC Current

1. Are the experimental values within the margin of error allowed by the uncertainty in the resistor value and noise from the system? How can the error be minimized in this experiment?

RC Current

2. Considering the sum voltages in the 5th column of the RC Voltages table under the RC Analysis tab, what can you conclude about Kirchhoff's Loop Law (Equation 2) for alternating current circuits?

1440 Questionnaire: Experiment 7: Kirchoff's Circuit Laws

Do not put your name on this page. Hand in this page separately.

1. TA name(s):

2. What one thing did you like best about this laboratory?

3. What one thing did you like least about this laboratory?

4. What one thing would you change in this laboratory?

5. What one thing would you leave the same?

Additional Comments?

Last/First Name (print): _____

PHYS 1440-Section _____ Username ID (xxx9999): _____

Experiment 8:

Magnetic Field Mapping and Earth's Magnetic Field

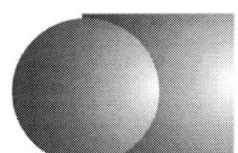

Pre-lab, Part A

Magnetic Field Mapping

1. Describe the orientation of magnetic field lines by drawing a bar magnet, labeling the poles, and drawing several lines indicating the direction of the forces.

2. How is the magnetic field strength **B** related to the density of the field lines?

Pre-lab, Part B
Earth's Magnetic Field

3. Describe the various ways in which the dip angle will be measured.

4. How many degrees is the Magnetic Field Sensor rotated through both the horizontal plane and the vertical plane?

Last/First Name (print):

Experiment 8:
Magnetic Field Mapping and Earth's Magnetic Field

PART A: Magnetic Field Mapping

Introduction

The purpose of this experiment is to visualize the magnetic field by using small compasses to trace magnetic field lines for a dipole, a repulsive dipole, and a quadrapole field.

Theory

Magnetic field lines are used to help visualize the magnetic field. There are some rules for drawing these lines.
- The lines begin on a north pole and terminate on a south pole. This is actually only true for the field external to the magnet. Inside the magnet, the lines complete full loops and point from south to north. See Figure 1.
- The magnetic field strength is directly proportional to the density of the field lines. This is really only true in three dimensions, but in 2-D drawings it is still true that the field is stronger where the lines are closer together.
- The magnetic field (**B**) is tangent to the field line at any point. Note that **B** is a vector.
- The field should have the same symmetry as the magnet configuration that produced it.
- The lines (of the total field) cannot cross (although since the compasses used to trace the lines are not infinitely small, places where the field is sharply curved can be hard to follow and lines may seem to cross.)

- The lines cannot stop or start in space. There is one exception here. On a symmetry axis there may be a place where the field is zero. Most lines must avoid such a place, but a line on the symmetry axis must stay on the symmetry axis. On either side of the zero, the field must either point toward the zero or away from it. Thus it appears as if two lines have either started or stopped at the zero. However, there are at most only a few symmetry axes, but an infinite number of field lines (we don't draw them all). This means the fraction of badly behaved lines is (small #)/infinity = 0, so we may ignore the poorly behaved lines.

A compass needle (small magnet) will align with the (average) **B** field at the region where the needle is as shown in Figure 1. Thus, we can use a small compass to trace the lines.

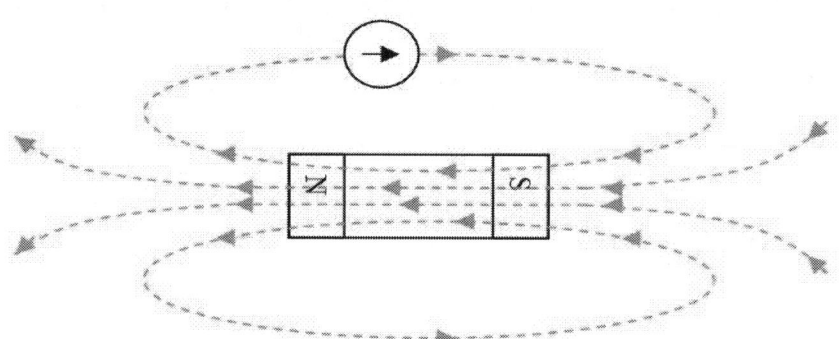

Figure 1: Dipole Field with Compass

Equipment

1	Plotting Compasses (Set of 20)	EM-8680
1	Bar Magnet Alnico (Set of 2)	EM-8620
	Paper & Pencil	
1	PASCO Capstone	UI-5400

Setup/Procedure A: Magnetic Field Mapping

Note 1: The PASCO EM-8620 Alnico magnet has a groove on the North end.
Note 2: If the laboratory tables have metal parts, they can become magnetized and add to the total magnetic fields you are mapping. You can check you table top with a compass to see if it is affected in the region where you intend to do your field plotting. Once a suitable location has been found, begin the experiment.

Attractive Dipole
1. Position the magnets so they are about halfway onto a piece of 8½ x 11 inch paper as shown in Figure 2. The north pole of one magnet should face the south pole of the other.
2. Draw a line around each magnet and label the poles.
3. Make a semi-circle of nine dots with one dot on the symmetry axis and four above the axis and four below as shown near the north end.
4. Position the compass so the south end of the compass points to one of the dots in the semi-circle (in Figure 2, I started at the third dot from the top).
5. Use a wooden pencil to make a dot on the paper where the north end of the compass needle points.
6. Move the compass so the south end points to the dot you just made and mark a new mark where the north end points.
7. Continue until you go off the paper or reach the south pole of the other magnet.
8. Fit a smooth curve to the dots. Don't forget that the field is a vector!
9. Now do the other eight points. Note: this should not require a lot of time. This is an imprecise exercise.

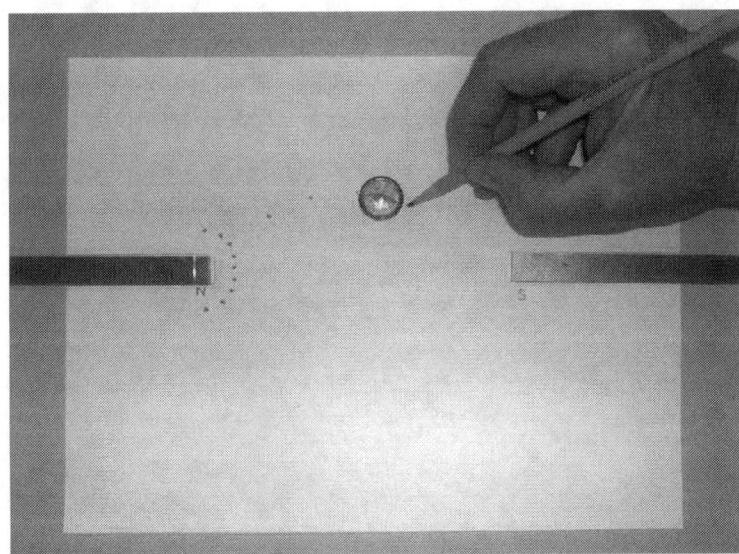

Figure 2: Attractive Dipole Set-up

Repulsive Dipole:
1. Flip the paper over to record the repulsive dipole data.
2. Set up as before, but let the two north ends face each other. See Figure 3.
3. Draw a semi-circle of nine dots around each north pole.
4. Trace all 18 lines.

Quadrapole:
1. Use a long piece of paper
2. Draw an arrow on the paper with the north direction indicated as shown in figure 4. With the magnets far away, use your compass to align the paper so the arrow points at magnetic north.
3. Put the magnet on the paper at a 45-degree angle as shown in figure 4.
4. Mark the position of the magnet and indicate which pole is north.
5. Put a semi-circle of nine dots around each pole.
6. Put two dots along the side of the magnet and two more across from it.
7. Trace all 22 lines from each dot..

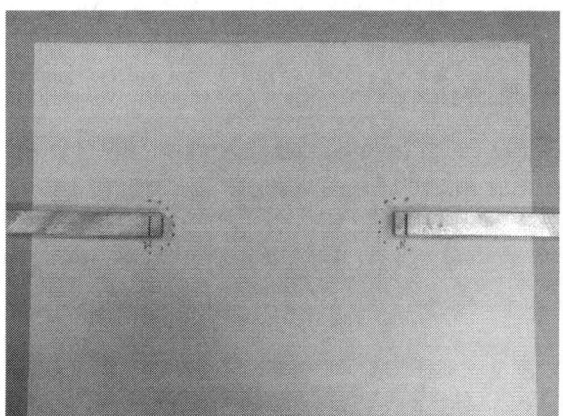

Figure 3: Repulsive Dipole

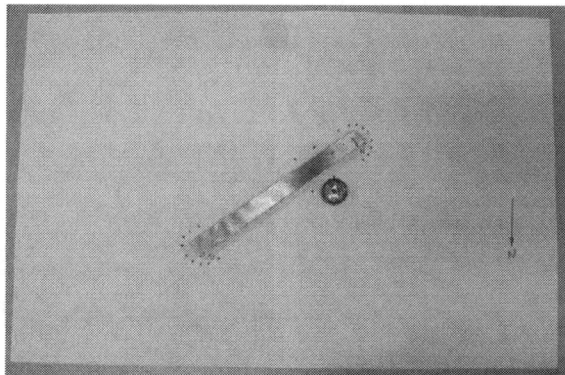

Figure 4: Quadrapole

Last/First Name (print): _____

PHYS 1440-Section _____ Username ID (xxx9999): _____

Analysis:

Discuss each of the three patterns briefly. Consider the six points discussed in the Theory section. Where is the field relatively strong? Are there any symmetry? Are there any zeroes? Do the lines cross? Do any lines start or stop in space?

PART B: Earth's Magnetic Field

Introduction

The magnitude and direction of the Earth's magnetic field are measured using a Magnetic Field Sensor mounted on a Rotary Motion Sensor. The Magnetic Field Sensor is rotated through 720 degrees in a horizontal plane and then 720 degrees in a vertical plane. The Rotary Motion Sensor will be turned by hand. This allows a determination of the horizontal component of the Earth's magnetic field, the total field and the dip angle. The Magnetic Field Sensor is zeroed using the Zero Gauss Chamber, the walls of which are made of a highly permeable material which redirects the magnetic field around the chamber.

Theory

The magnitude of the Earth's field varies over the surface of the Earth. The horizontal component of the Earth's magnetic field points toward the Magnetic North Pole (which must therefore have a South polarity). The north end of a compass needle is attracted to the south end of the Earth's magnetic field. The magnetic pole close to the geographic north pole, "Magnetic North" is actually a magnetic south pole.

The total magnetic field points at an angle from the horizontal. This angle (θ) is called the dip angle. An example for the Northern hemisphere is shown in figures 1 and 2.

$$\cos \theta = B_{Horizontal}/B_{Total} \qquad \textbf{Equation 1}$$

The Magnetic Field Sensor detects the component of the magnetic field in a direction that is parallel to the clear probe on the sensor. If we rotate the sensor in a horizontal plane, the sensor will detect the component of $B_{Horizontal}$

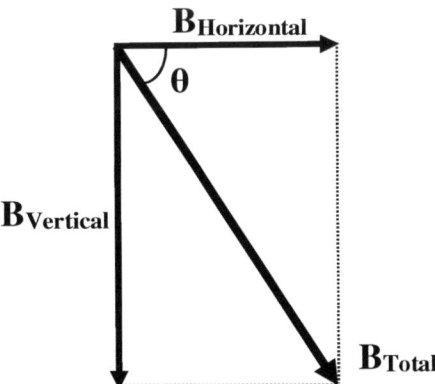

Figure 1: Components of the Magnetic Field (Northern Hemisphere)

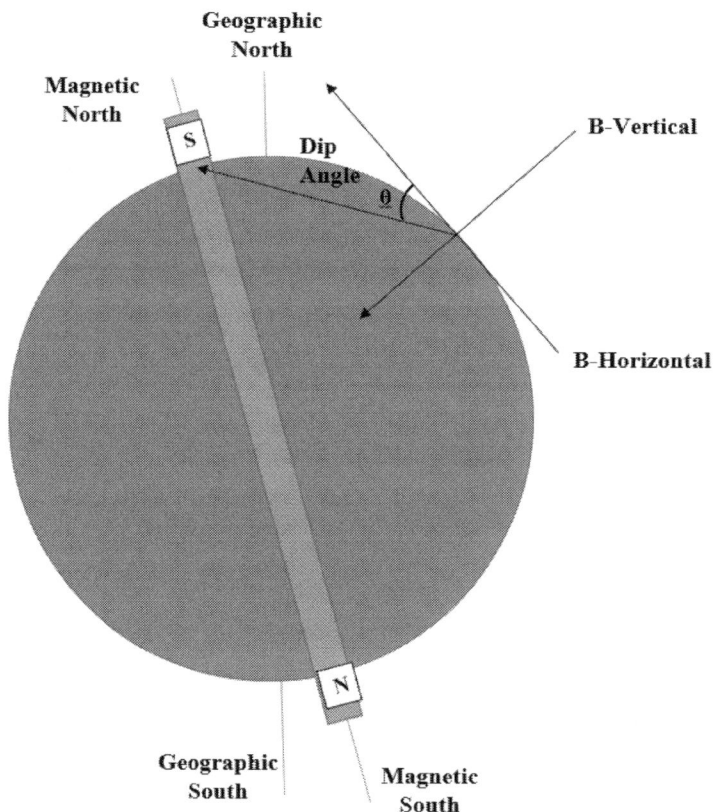

Figure 2: Dip angle with respect to the Earth

Equipment

1	2-Axis Magnetic Field Sensor	PS-2162
1	Zero Gauss Chamber	EM-8652
1	Rotary Motion Sensor	PS-2120
1	Dip Needle	SF-8619
1	Large Rod Base	ME-8735
1	90 cm Stainless Steel Rod	ME-8738
1	850 Universal Interface	UI-5000
1	PASCO Capstone	UI-5400

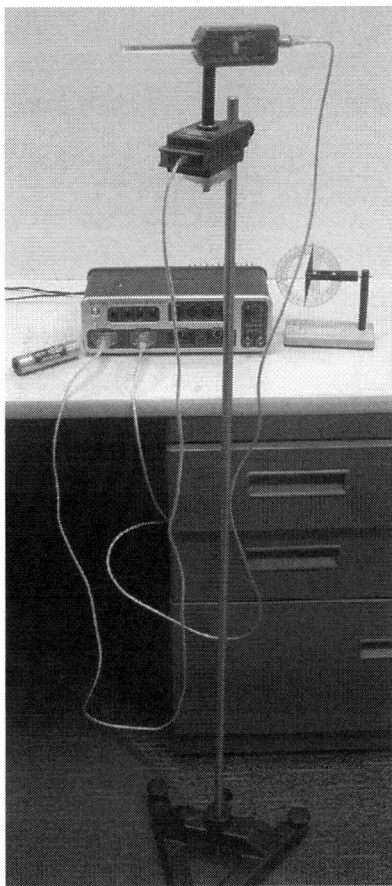

Figure 3: Setup

Setup: Earth's Magnetic Field

NOTE: During this experiment, keep the apparatus away from all sources of magnetic fields (electrical, computers, computer interface, bar magnets). Also keep away from all ferromagnetic materials (iron, steel chairs and tables). This is essential for good results since the Earth's Field is orders of magnitude smaller than the field near a refrigerator magnet.

1. Assemble the system as shown in Figure 3. Note that the raised key inside the Magnetic Field Sensor handle slides into the notch on the shaft of the Rotary Motion Sensor.
2. Use a nonmagnetic stainless steel rod in the rod holder.
3. Attach the Rotary Motion Sensor as high as possible on the stainless steel rod to keep the Magnetic Field Sensor as far away from the rod as possible.
4. Plug the Rotary Motion Sensor and the Magnetic Field Sensor into any two *PASPORT* inputs on the 850 Universal Interface.

Procedure B: Earth's Magnetic Field

Horizontal Component of the Magnetic Field of the Earth

1. Use a small compass to orient the front of the probe to face north.
2. Level the top of the case of the Rotary Motion Sensor along its long axis.
3. Rotate the Magnetic Field Sensor so the length of the probe is pointing east or west.
4. Slip the Zero Gauss Chamber over the Magnetic Field Sensor probe.
5. Press the Tare button on the Magnetic Field Sensor.
6. Release the Tare button and remove the Zero Gauss Chamber. The horizontal component of the magnetic field is zero at an angle of 90 degrees from north in the horizontal plane. Pushing the Tare button here zeros the sensor at this point. However, the magnetic field strength is only a few hundredths of a mT and any noise is of the same order of magnitude. Instead of seeing zero when perpendicular to the field, the measured field will be shifted vertically by this zero error depending on which reading it uses for zero. This will not affect the experiment.
7. Align the Magnetic Field Sensor with the long axis of the Rotary Motion Sensor so it points due north.
8. Click RECORD.
9. **Slowly and steadily** rotate the Rotary Motion Sensor pulley through two and one quarter revolutions clockwise.
10. Click STOP. It will help if someone holds the cable out of the way. If the angles on the graph are negative it is okay. Positive angles can be obtained by rotating in the opposite direction.
11. The field should have its most negative value near 0^0, 360^0, and 720^0. If not, the system needs to be realigned, and steps 1-10 need to be repeated.
12. Click on Data Summary at the left of the screen. Double click on this run and re-label it "Horizontal 1".
13. Repeat steps 7-12 twice more, labeling the runs "Horizontal 2" and "Horizontal 3".

14. Sketch the data in graph 1.

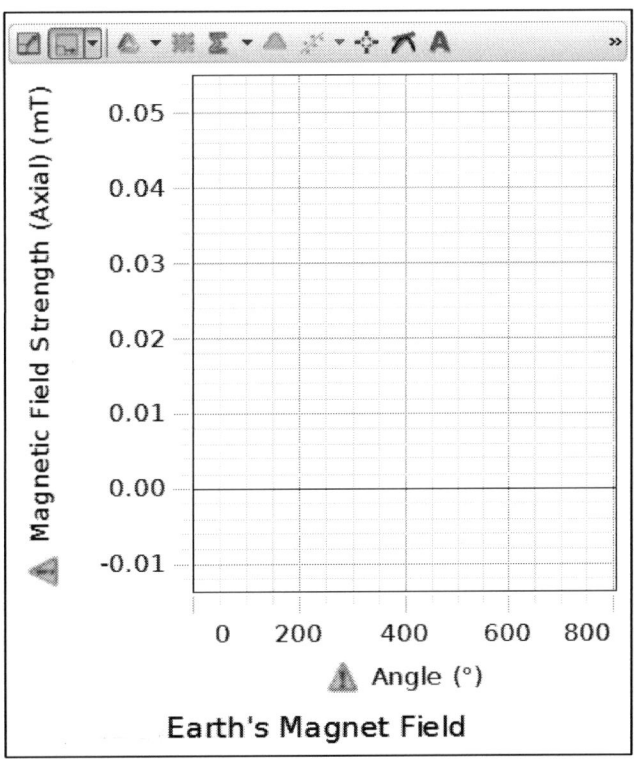

Graph 1: Horizontal Magnetic Field Strength vs. Angle

Total Magnetic Field of the Earth

1. To allow the Magnetic Field Sensor to rotate in a vertical circle, remove the Rotary Motion Sensor from the rod.
2. Rotate it 90^0.
3. Reattach it to the rod.
4. Align the Rotary Motion Sensor north using a compass.
5. Point the Magnetic Field Sensor probe horizontally.
6. Click RECORD with the Magnetic Field Sensor still horizontal.
7. **Slowly and steadily** rotate the Rotary Motion Sensor pulley through two-and-one-quarter revolutions in a direction so probe turns **downward**.
8. Click STOP.
9. Click on Data Summary at the left of the screen.
10. Double click on this run and re-label it "Total 1".
11. Repeat steps 5-10 twice more. Label the runs "Total 2" and Total 3".
12. Hold the Dip Needle in its horizontal position so it is level. Align it so the needle points to the 270^0 mark.
13. Rotate the fork 90^0 so the needle pivots in a vertical plane.
14. Allow the needle to come to rest and read the number of degrees it is below the horizontal plane (270^0).
15. Record the value in line 5 of the table 2.
16. Sketch the data in graph 2.

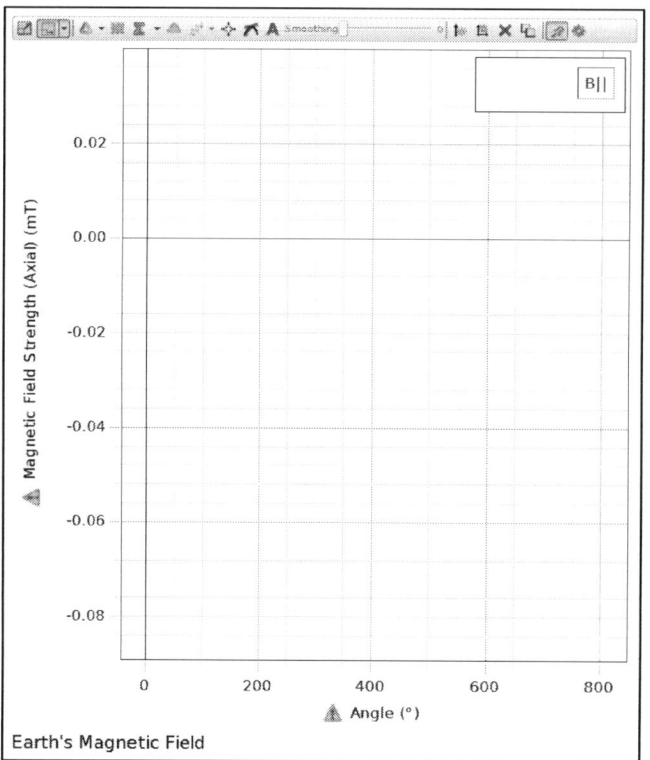

Graph 2: Vertical Magnetic Field Strength vs. Angle

Last/First Name (print): _____

PHYS 1440-Section _____ Username ID (xxx9999): _____

Analysis:

1. Click the black triangle by the Run Select icon on the graph toolbar and select "Horizontal 1".
2. Click the Scale to Fit icon on the graph toolbar.
3. Click the Selection icon on the graph toolbar and drag the handles on the Selection box to highlight the data from just after you began taking data to just before you ended. Stay away from the endpoints.
4. Click the black triangle by the Curve Fit icon and select Sine The sine curve on the screen may match your data well, but the computer may struggle matching such noisy data. If the curve is clearly wrong, try decreasing the width of the Selection box by moving the beginning and ending handles. Move the selection box around until a sine curve fit snaps to the data.
5. Record the absolute value of the amplitude of the sine curve (A in the Sine box) in the "Horizontal" row, and "Run 1" column of table 1. This is the value for $B_{Horizontal}$ or B_{Total} for the vertical runs.
6. Repeat steps 1-4 for the other two horizontal runs.
7. Repeat steps 1-4 for the three vertical runs, recording the amplitudes in the Total row.
8. Note that for both the Horizontal field and the Total Field, the computer calculates the average value and the uncertainty for the fields from your three values.

Table 1: Earth's Magnetic Field

	Field	Run 1 (mT)	Run 2 (mT)	Run 3 (mT)	Average (mT)	Uncertainty (mT)
1	Horizontal					
2	Total					

The Dip Angle
1. Click on the black triangle by the Run Select icon on the graph toolbar and select "Total 1".
2. Click on the Selection icon and drag the handles on the Selection box to highlight data from just after you started the run to around 400^0 (one full cycle and the beginning of the second cycle).
3. Click the Scale to Fit icon.
4. Click the black triangle by the Curve Fit icon and select Sine. As before, the selection box handles may have to be adjusted to fit the data.
5. Estimate the angle at which the first minimum occurred and record the value in line 1 of the Dip Angle table.
6. Repeat for the other two total runs, entering the values in lines 2 and 3 of the table.
7. Calculate the average of the three values and enter it in line 4.
8. Calculate the dip angle using equation 1 and the average values of $B_{Horizontal}$ and the total magnetic field. Enter the value in line 6 of the table 2.

Table 2: Dip Angle

	Angle From	Dip Angle (degrees)
1	Graph Total 1	
2	Graph Total 2	
3	Graph Total 3	
4	Graph Average	
5	Dip Needle	
6	Equation 1	

Do the different methods of finding the Dip Angle agree?

1440 Questionnaire: Experiment 8: Magnetic Field Mapping and Earth's Magnetic Field

Do not put your name on this page. Hand in this page separately.

1. TA name(s):

2. What one thing did you like best about this laboratory?

3. What one thing did you like least about this laboratory?

4. What one thing would you change in this laboratory?

5. What one thing would you leave the same?

Additional Comments?

Last/First Name (print): _____

PHYS 1440-Section _____ Username ID (xxx9999): _____

Experiment 9:

Magnetic Field in a Current Carrying Coil

Pre-lab

1. Calculate the current in a long solenoid having a magnetic field strength of 3.5×10^{-5} T and 300 loops per unit length.

2. When will the magnetic field strength equation for a long solenoid fail?

3. How can we distinguish long and short solenoids? In other words, what is the condition that actually makes a long solenoid considered to be 'long'?

4. Why won't equation 1 or 2 work for the short solenoid used in this experiment?

Last/First Name (print): _____

Experiment 9:
Magnetic Field in a Current Carrying Coil

Introduction

In this experiment we are comparing changes in axial and radial magnetic field components as the position of a magnetic field sensor is moved through a current carrying coil. The position is recorded by a string attached to the Magnetic Field Sensor that passes over the Rotary Motion Sensor pulley to a hanging mass.

Theory

Axial and Radial Field

The Magnetic Field Sensor is able to record the axial and radial components of the magnetic field as shown in Figure 1. The axial component of the magnetic field is the field that points in the direction of the solenoids vertical axis. The radial component of the magnetic field is the portion of the field that points outward from the center to the walls of the solenoid and beyond. Along with the magnitude of each component, the sensor detects the direction of each field component. The direction is relative to the orientation of the sensor and is denoted with positive and negative values.

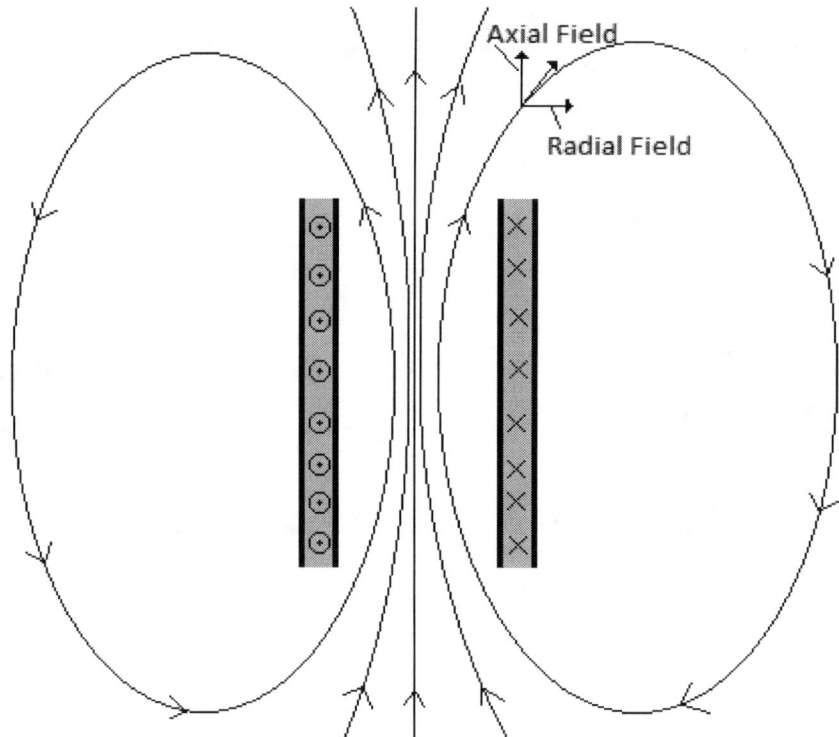

Figure 1. Visualization of magnetic field through the solenoid

Single Coil

For a coil of wire of negligible length, as shown in Figure 2, having radius R and N turns of wire, the magnetic field along the perpendicular axis through the center of the coil is given by

$$B = \frac{\mu_0 N I R^2}{2(x^2 + R^2)^{3/2}}$$

Equation 1

where $\mu_0 = 4\pi \times 10^{-7} \, T \cdot \frac{m}{A}$ is the permittivity of free space, N is the number of turns, R is the radius of the coil, I is the current through the coil, and x is the distance from the center of the coil.

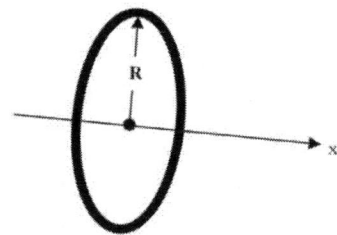

Figure 2: Single coil of Radius R

Long Solenoid (n coils)

For a long solenoid with n turns per unit length, the magnetic field is

$$B = \mu_0 n I \qquad \textbf{Equation 2}$$

where $\mu_0 = 4\pi \times 10^{-7} \, T \cdot \frac{m}{A}$, n is the number of turns per unit length, and I is the current through the solenoid.

The direction of the field is straight down the axis of the solenoid. To be considered "long", the length of the coil must be much longer that the diameter of the coil as shown in Figure 3. In addition, Equation 2 will begin to fail when the ends of the solenoid are approached and the magnetic field strength will begin to decrease.

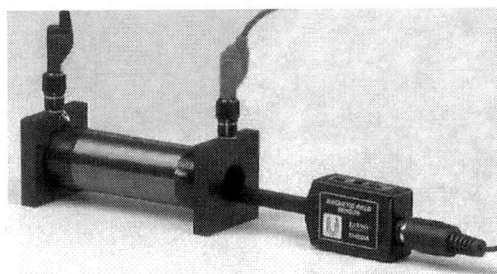

Figure 3: Long Solenoid

Short Solenoid

For the short solenoid on the AC/DC Electronics Laboratory, neither Equation 1 nor Equation 2 is correct since the coil is too long for Equation 1 to work and too short to use Equation 2. However, both equations tell us something about the behavior of the magnetic field. Both equations should yield upper bound values for the magnetic field at the exact center of the coil, since in the first case squeezing the 600 turns into zero length would make all of the coils closer and in the second case adding more coils would clearly increase the field. The specifications for the coil used in this experiment can be found in Table 1.

Table 1: Specifications for the Short Solenoid

Number of Turns	600 turns
Radius	1.5 cm
Length	2.5 cm

Equipment

1	AC/DC Electronics Laboratory	EM-8656
1	Rotary Motion Sensor	PS-2120
1	2-Axis Magnetic Field Sensor	PS-2162
1	Short Patch Cords (set of 8)	SE-7123
1	Large Rod Base	ME-8735
1	90 cm Rod, Stainless Steel	ME-8738
1	Mass and Hanger Set	ME-8979
1	Black Thread	ME-9875
1	850 Universal Interface	UI-5000
1	PASCO Capstone	UI-5400

Setup

1. Connect red and black patch cables between the red and black jacks for Output 1 on the Universal Interface and the Electronics Laboratory.
2. Connect wires between the spring clips attached Electronics Laboratory inputs and the clips on either side of the yellow coil.
3. Attach the 90 cm rod to the base and position the base by the Electronics Laboratory as shown in Figure 4.
4. Attach the Rotary Motion Sensor (RMS) to the rod as in Figure 3. Cut a piece of thread about 1 m long. Tie one end of the thread around the strain relief in the cable attached to the Magnetic Field Sensor so it catches in one of the grooves as shown in Figure 4. Pass the other end of the thread over the large step of the RMS pulley and attach a mass holder carrying an additional 60 g of mass.
5. Adjust the positions so the thread between the RMS and the Magnetic Field Sensor is vertical when the sensor is centered on the coil. Attach the sensor handle to the Magnetic Field sensor. Adjust the positions so that while holding the end of the sensor handle against the 90 cm rod, you can move the sensor up and down along the magnetic field axis.
6. Plug the Magnetic Field Sensor and the Rotary Motion Sensor into any two of the *PasPort* inputs on the 850 Universal Interface.

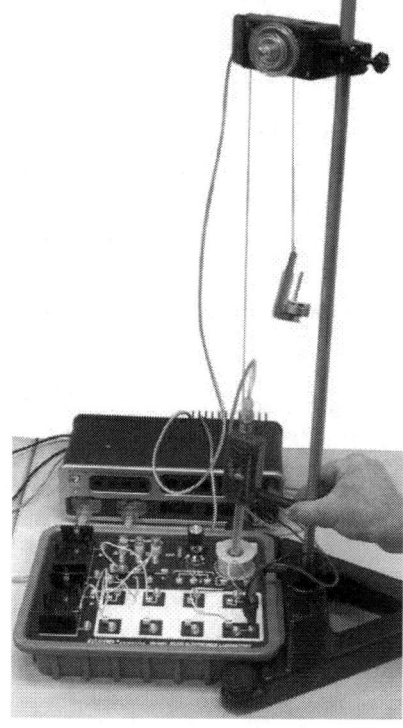

Figure 4

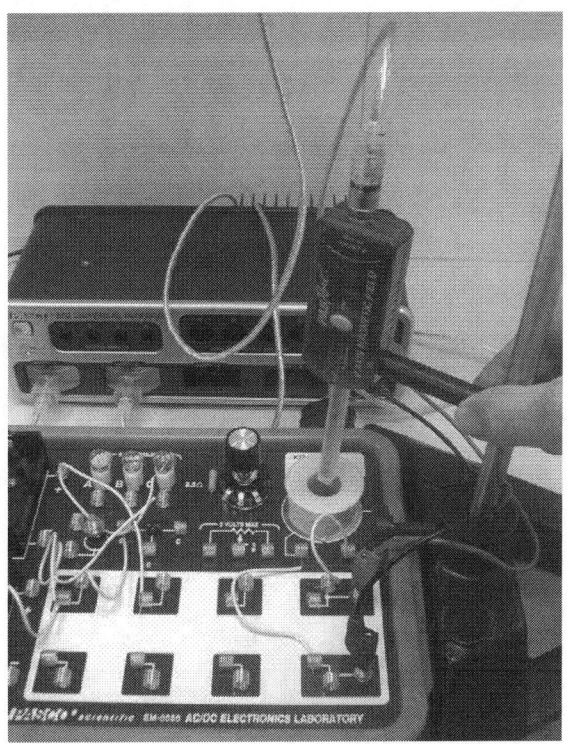

Figure 5: Circuit Board Connections

Last/First Name (print): _____

PHYS 1440-Section _____ Username ID (xxx9999): _____

Procedure

1. Open the signal generator at the left of the screen and turn the DC power off
2. Put the Magnetic Field Sensor probe into the coil so the sensor is at the center of the coil and the handle is facing north. The vertical position of the sensor is indicated by the white dot on the side of the probe away from the labels on the sensor body.
3. Press the Tare button (green button) to zero the sensor.
4. Open Data Summary.
5. Click Rotary Motion Sensor properties, and click "Zero Sensor Now" then OK.
6. Open the Signal Generator. Set Output 1 to a DC Voltage of 5 V.
7. Click Auto, then click the signal generator again to close it.
8. Insert the magnetic sensor probe through the solenoid as far as possible.
9. Click RECORD.
10. Move the probe slowly out of the solenoid keeping it centered with the handle facing north. Continue until the bottom of the sensor is 10 cm above the coil.
11. Click STOP. Record the current in Table 2.
12. If the curve is not symmetric about Position = 0 cm, repeat step 5. Small deviations from the center less than 3 cm are acceptable.
13. Open Data Summary and re-label this run as "Center". Record the Current in Table 2.
14. Select the coordinate tool. Right click the tool and change significant figures to 5. Measure the peak amplitude and record in Table 2.
15. Repeat steps 10-14 with the probe shifted so that the probe is touching the east edge of the coil. Keep the handle pointed north. Label this run as "East".
16. Repeat steps 10-14 with the probe against the west edge of the coil. Keep the handle pointed north. Label as "West".
17. Reverse the red and black patch cords so the current through the coil is reversed. Repeat step 15 and label it "East Reversed".
18. Calculate the bound values for this solenoid by calculating the field strength for both a short solenoid and a long solenoid using equation 1 and 2. Use the center current measured during the center run. Record these values in Table 3. The calculated value will be in Tesla (T). Convert this value to mT and record in Table 3. For Equation 1, $x = 0$.

Table 2: Current through Coil

Trial	Coil Current (amps)	Axial Peak Amplitude (mT)
Center		
East		
West		
East Reversed		

Table 3: Bound Values

Model	Magnetic Field Strength (mT)
Short Solenoid (equation 1)	
Long Solenoid (equation 2)	

Last/First Name (print): _____

PHYS 1440-Section _____ Username ID (xxx9999): _____

Analysis

Axial Field Model:
1. How do the first three runs compare with each other? What does this show about the axial field strength across the coil (moving perpendicular to the axis of the coil)?

2. Does your data indicate that the field is upward or downward? Using the right hand rule, is the current in the coil clockwise or counterclockwise viewed from above?

3. Why is the "East Reversed" run different from the other three?

4. How well did the estimates given by equations 1 & 2 work out? Is the measured value of the solenoid closer to a long solenoid, or a short solenoid? Use the Peak Amplitude found in Table 3 to answer this question.

5. Sketch the axial field data.

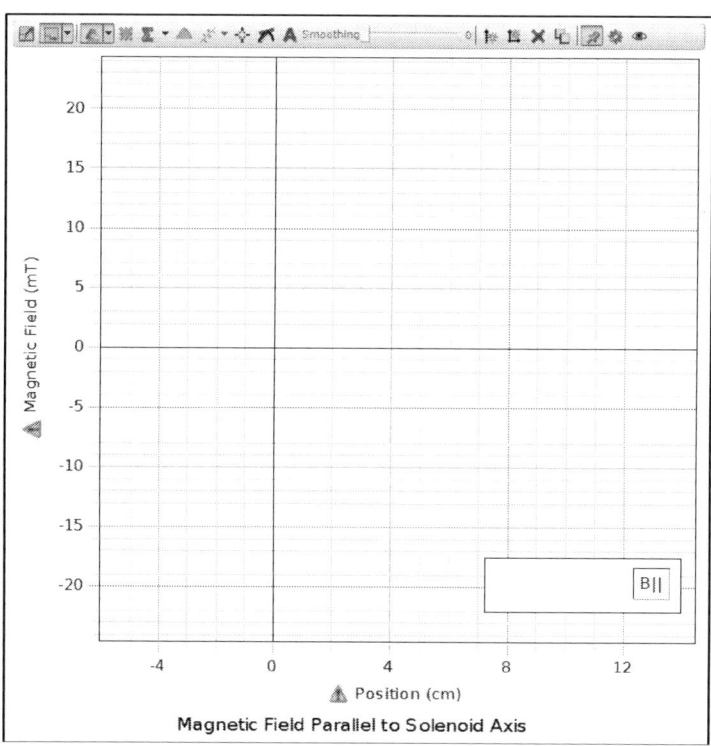

Graph 1: Axial Field Graph

Last/First Name (print): _____

PHYS 1440-Section _____ Username ID (xxx9999): _____

Radial Field Model:
1. On the axis, there should be no perpendicular field due to the symmetry of the system. There is probably a small radial field due to failure to get the sensor right on the axis. Does the field appear small for the "Center" run? Click the black triangle by the Run Select icon on the graph toolbar. Click "Center" to de-select it. The "East" and "West" runs should still show.

2. Note that the positive direction for the radial field is coming out of the side of the sensor with the labels on it. Explain why the "East" radial field looks the way it does.

3. Explain why the "West" radial field look the way it does.

4. Click on the black triangle by the Run Select icon and select "East Reversed" and click on "East" to de-select it. Explain why "West" and "East Reversed" appear different.

5. Sketch the radial field data.

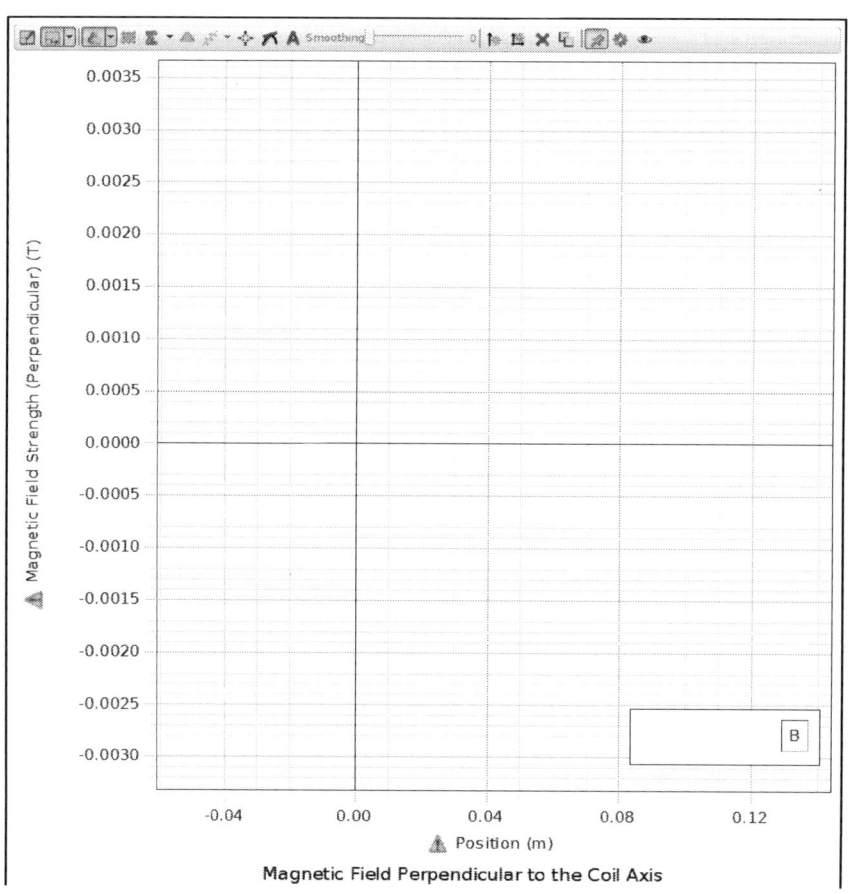

Graph 2: Radial Field

1440 Questionnaire: Experiment 9: Magnet Field of a Current Carrying Coil

Do not put your name on this page. Hand in this page separately.

1. TA name(s):

2. What one thing did you like best about this laboratory?

3. What one thing did you like least about this laboratory?

4. What one thing would you change in this laboratory?

5. What one thing would you leave the same?

Additional Comments?

Last/First Name (print): _____

PHYS 1440-Section _____ Username ID (xxx9999): _____

Experiment 10:
Induction-Magnet Through a Coil

Pre-lab

1. In this experiment, how will the magnetic flux in a coil of wire be changed?

2. When the magnetic flux through a coil of wire changes, what is generated between the ends of the coil? Which equation predicts this?

3. Define all quantities in Faraday's Law.

4. How can magnetic flux be thought of intuitively?

Last/First Name (print): _____

Experiment 10:
Induction-Magnet Through a Coil

Introduction

The purpose of this experiment is to examine Faraday's Law of Induction. A magnet will be dropped through a coil and the voltage across the coil will be graphed as a function of time. The total flux of the magnet moving into the coil will be compared to the total flux of the magnet moving out of the coil.

Theory

When the magnetic flux through a coil of wire changes (as in a magnet falling through a coil of wire in Figure 1), an EMF (ε) is generated between the ends of the coil given by Faraday's Law:

$$\varepsilon = -N \, (d\Phi/dt) \quad \textbf{Equation 1}$$

where N is the number of turns in the coil and $d\Phi/dt$ is the time rate of change of the magnetic flux, Φ, or the derivative of the magnetic flux with respect to time. The magnetic flux may be thought of as the number of magnetic field lines (Figure 1) passing through the coil. Integration of Equation 1 yields:

$$\int \varepsilon \, dt = -N\Delta\Phi = \text{the area under the curve on an } \varepsilon \text{ vs t graph} \quad \textbf{Equation 2}$$

where $\Delta\Phi$ is the total change in flux (total number of field lines).

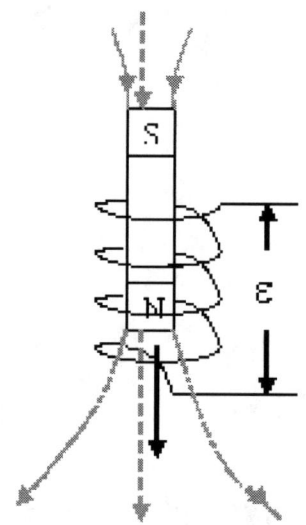

Figure 1: Falling Magnet

Equipment

1	AC/DC Electronics Laboratory	EM-8656
1	Voltage Sensor	CI-6503
1	Bar Magnet Alnico (set of 2)	EM-8620
1	No-Bounce Pad	SE-7347
1	850 Universal Interface	UI-5000
1	PASCO Capstone	UI-5400

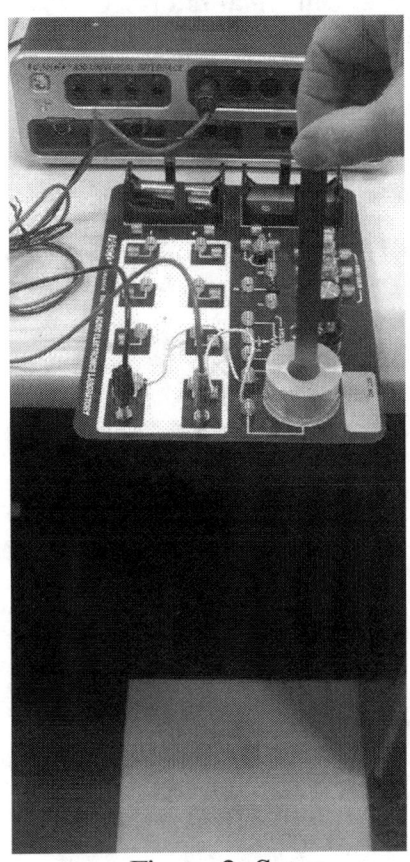

Figure 2: Setup

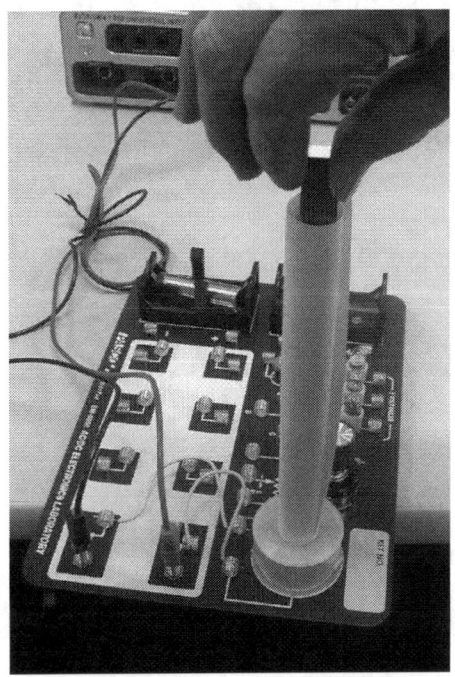

Figure 3: Paper Roll Guide

Setup

NOTE: During this experiment, keep the apparatus away from all sources of magnetic fields (electrical, computers, computer interface, bar magnets). Also keep away from all ferromagnetic materials (iron, steel chairs and tables). This is essential for good results since the Earth's Field is orders of magnitude smaller than the field near a refrigerator magnet.

1. Connect wire jumpers from the coil connector springs to the banana inputs for the Electronics Laboratory as shown in Figure 2.
2. Connect red and black ends of the Voltage Sensor to the banana inputs on the Electronics Laboratory.
3. Plug the Voltage Sensor into Analog Channel A on the 850 Universal Interface.
4. Position the Electronics Laboratory so it hangs over the desk enough so a magnet can be dropped through the coil. Place something heavy in one of the battery holders to prevent the Electronics Laboratory from tipping.
5. Place the No-Bounce pad on the floor where the magnet may hit. Roll up a piece of paper tightly to make a tube 8.5 inches long.
6. Insert the paper tube into the coil as in Figure 3. This will guide the falling magnet so it hits the hole. Note where the magnet must be held so its lower end is about 5 cm above the coil.

Procedure

Note: the PASCO EM-8620 Alnico magnet has a groove near the North end.

1. Hold one magnet so its north end is downward and is in the paper tube about 5 cm above the coil.
2. Click open the Recording Conditions on the bottom tool bar.
3. Click on Stop Condition. Set Condition Type to TIME BASED.
4. Set Record Time to 2.0 s.
5. Click OK.
6. Click RECORD.
7. Drop the magnet. **Remember to catch the magnet before it hits the floor! If it is not caught the magnets may break!** Data collection will start and stop automatically. The graph should look like Figure 4. If you only see an "up pulse", reverse the connections on the coil by reversing the red and black banana plugs from the voltage sensor.
8. Click open Data Summary at the left of the page.
9. Re-label this run as "N down".
10. Click on Recording Conditions at the bottom of the page.
11. Change the Start Condition to "Falls Below" and the Value to -0.0100.
12. Click OK.
13. Repeat steps 6-8 with the south end down.
14. Label it "S down".
15. Click on Recording Conditions at the bottom of the page.
16. Change the Start Condition to "Rises Above" and the Value to 0.0100.
17. Click OK.
18. Repeat steps 6-8 with the north end down but held as high as possible in the paper tube.
19. Label this run "N down hi".
20. Remove the paper tube.
21. Stick the two magnets together (north to south).
22. Hold them so one end is already in the top of the coil (they will just barely fit).
23. Repeat 6-8 and label "N to S".
24. Tape the two magnets together so the north poles are together and the south poles are together.
25. Hold them so the north end is already in the top of the coil (they will just barely fit).
26. Repeat step 6-8 and label "N to N".

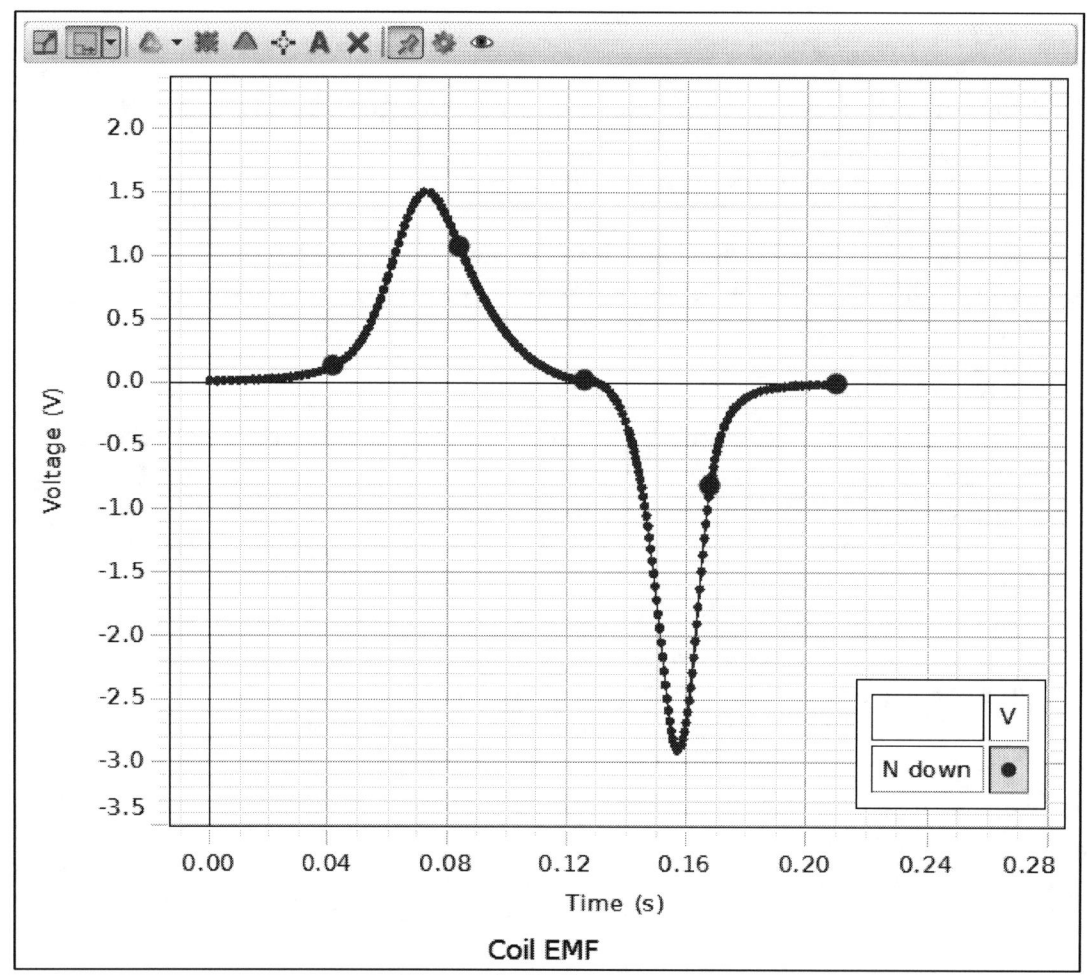
Figure 4: Coil EMF

Last/First Name (print): _____

PHYS 1440-Section _____ Username ID (xxx9999): _____

Analysis

1. Click on the black triangle by the Run Select icon in the graph toolbar.
2. Select "N down".
3. Click on the Scale-to-Fit icon.
4. Click on the Properties icon in the graph toolbar.
5. Show Run Symbols and Show Data Points should not be selected (so you can see the line better). If they are selected, de-select them in both the Active Data Appearance box and the Future Data Appearance box.
6. Click OK.
7. Click on the Selection icon in the graph toolbar.
8. Adjust the handles on the selection box to select the data time = 0 to the point where V = 0 as the curve crosses the horizontal axis. You should have selected all the positive data.
9. Click the Area icon. It will calculate the area under the curve. Recall from Theory that this is the change in the magnetic flux through the coil. If the Area box does not show three significant figures, click on Data Summary. Click on the Voltage Properties icon (not the Voltage Sensor Properties) and change to three significant figure under the Number Format. Enter area of the first pulse in the First Pulse column of Table 1 on the N down line.
10. Move the selection box to select all the area for the second pulse.
11. Enter the value in the Second Pulse column of Table 1.
12. Repeat for the "S down", "N down hi", "N to S" and "N to N" data sets.
13. Sketch the data for each orientation in graph 1.

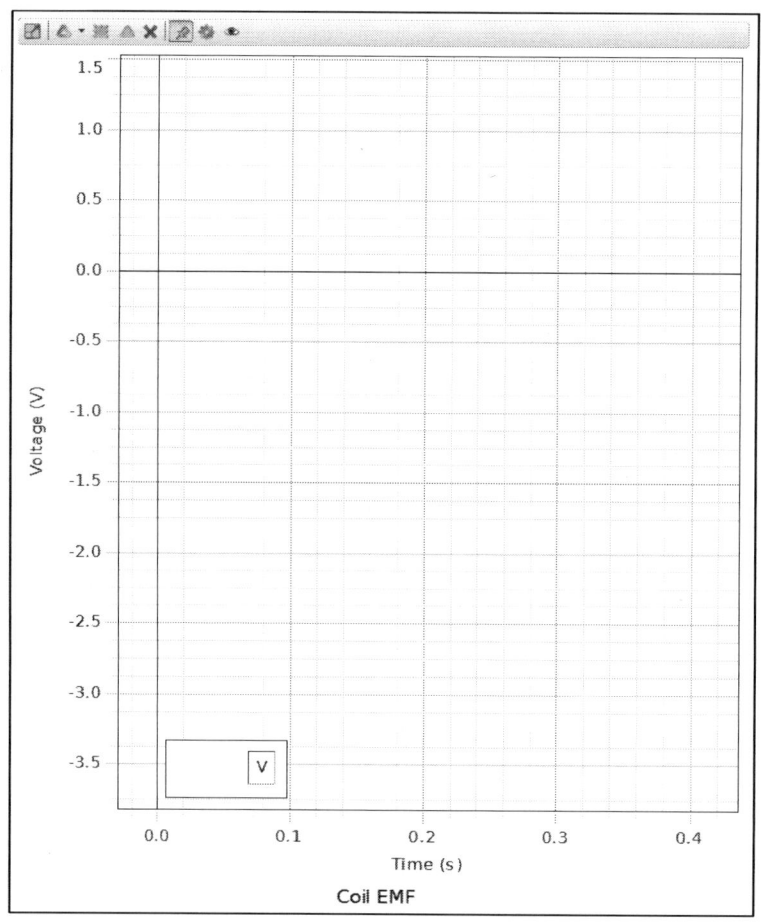

Graph 1: Voltage vs. Time

Table 1: Change in Magnetic Flux = N ΔΦ

No.	System	First Pulse (Vs)	Second Pulse (Vs)
1	N down		
2	S down		
3	N down hi		
4	N to S		
5	N to N		

Last/First Name (print): _____

PHYS 1440-Section _____   Username ID (xxx9999): _____

Conclusions:

1. On the graph, select the "N down" run. Why is the peak voltage higher on the 2nd pulse than on the 1st pulse?

2. Why is one pulse up and the other pulse down?

3. Explain why the value for the magnitudes of the change in magnetic flux for the "N down", "S down", and "N down hi" are all essentially equal.

4. Select both the "N down" and "S down" runs. In terms of the magnetic flux, explain why they are reversed.

5. Select the "N down" and "N down hi" runs. Explain why they are different. Why doesn't this change the area under the curve?

6. Select the "N down" and "N to S" data. Explain the difference.

7. Explain why the "N to N" data in Table 1 are different from the other cases.

1440 Questionnaire: Experiment 10: Induction-Magnet Through a Coil

Do not put your name on this page. Hand in this page separately.

1. TA name(s):

2. What one thing did you like best about this laboratory?

3. What one thing did you like least about this laboratory?

4. What one thing would you change in this laboratory?

5. What one thing would you leave the same?

Additional Comments?

Last/First Name (print): _____

PHYS 1440-Section _____ Username ID (xxx9999): _____

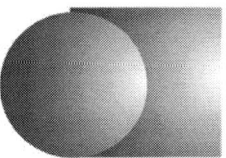

Experiment 11:
LRC Circuit-Resonance

Pre-lab

1. Impedance, Z, in an AC LRC circuit is the DC analog of what?

2. The inductor, capacitor, and resistor obey the AC analogs of what law? Give the three relevant equations from the theory section.

3. How is impedance related to inductive reactance and capacitive reactance? An equation is sufficient.

4. When will the impedance be equal to the resistance? Is this minimum or maximum impedance?

Last/First Name (print):

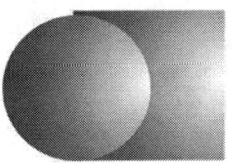

Experiment 11:
LRC Circuit-Resonance

Introduction

The resonance of a series LRC circuit is examined as a function of applied frequency and the effects of changing the values of the resistance, inductance, and capacitance. The phase difference between the applied voltage and the current is measured below resonance, at resonance, and above resonance.

Theory

An inductor, a capacitor, and a resistor are connected in series with a sine wave generator. Since it is a series circuit, the current will be common to all the components and given by

$$I = I_{max}cos(\omega t)$$ **Equation 1**

The voltage across the resistor will be in phase with the current, but the voltage across the inductor leads the current by 90^0 and the voltage across the capacitor lags by 90^0. Adding the three EMF voltages, **E**, results in a total voltage that varies sinusoidally, but has a phase shift φ with respect to the current and is given by

$$\mathbf{E} = \mathbf{E}_{max}cos(\omega t + \varphi)$$ **Equation 2**

The three components obey the AC analogs of Ohm's Law: $E_R = IR$, $E_{L\,max} = I_{max}X_L$, $E_{C\,max} = I_{max}X_C$, where X_L and X_C are the AC analogs of resistance called the inductive reactance and the capacitive reactance. The maximum current and total voltage are then related by

$$E_{max} = I_{max}Z \qquad \textbf{Equation 3}$$

where Z is called the impedance and is the AC analog of resistance for the entire circuit. The phase shift is related to the other variables by

$$\tan \varphi = \frac{X_L - X_C}{R} \qquad \textbf{Equation 4}$$

Since the voltage across the resistor is in phase with the current, the phase of the current can be measured by measuring the phase of the voltage across the resistor.

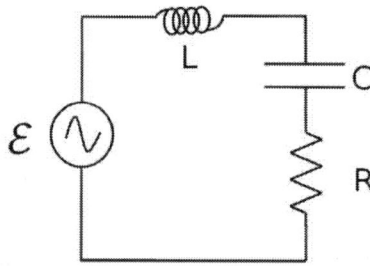

The impedance Z is given by

$$Z = \sqrt{R^2 + (X_L - X_C)^2} \qquad \textbf{Equation 5}$$

where the capacitive reactance is

$$X_C = 1/\omega C \qquad \textbf{Equation 6}$$

and the inductive reactance is

$$X_L = \omega L \qquad \textbf{Equation 7}$$

At resonance, the current is maximum and thus the impedance is at its minimum. The minimum impedance (equation 5) occurs when $X_L = X_C$ yielding $Z = R$. Setting equation 6 equal to equation 7 yields the resonant frequency:

$$\omega_{res} L = 1/\omega_{res} C \qquad \textbf{Equation 8}$$

$$\omega_{res} = \frac{1}{\sqrt{LC}} \qquad \textbf{Equation 9}$$

Equipment

1	Resistor/Capacitor/Inductor Network	UI-5210
3	Voltage Sensor	UI-5100
1	BNC-to-Banana Cord for 850 Output	UI-5119
1	Patch Cords (Set of 5)	SE-7123
1	PASCO Capstone	UI-5400
1	850 Universal Interface	UI-5000

Setup:

1. Connect the BNC-to-Banana cord to the #2 Signal Generator and connect the red cord to one end of the 2.5 mH inductor on the circuit board.
2. Connect the other end of the inductor to the 560 pF capacitor in series
3. Connect the other end of the capacitor to the 1.0 kΩ resistor in series.
4. Connect the black cord to the open end of the resistor.
5. Connect a Voltage Sensor to Channel A on the 850 interface and attach the leads across the resistor, making sure the black cable from the voltage sensor is connected to the grounded side of the resistor.
6. Connect a Voltage Sensor to Channel B on the 850 interface and attach the leads across the leads of the Output #2 cable, making sure the black cable from the voltage sensor is connected to the black side of the signal generator.
7. Open the Signal Generator 850 Output 2 and choose the Sine Wave at a frequency of 10,000 Hz and an amplitude of 7 V. Leave the output on AUTO.

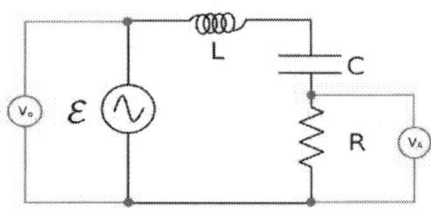

Figure 1: Series LRC Circuit

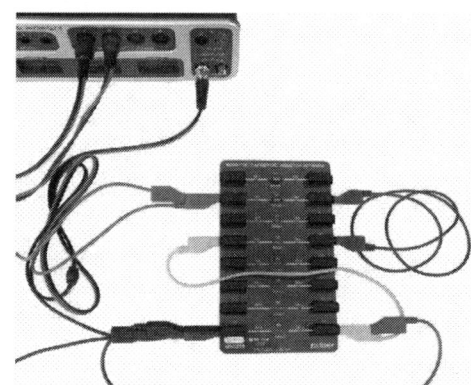

Figure 2: LRC Circuit with Sensors

Last/First Name (print): _____

PHYS 1440-Section _____ Username ID (xxx9999): _____

Procedure: Plotting the Resonance Curve

For this procedure the frequency of the applied voltage will be varied. The circuit response (current) will be recorded. This response is recorded by measuring the voltage across the resistor, because the current is in phase and proportional to it.

One further complication is that you must divide the resistance voltage by the output (of the 850) voltage to account for any changes in the output voltage. This works because if the output voltage doubles, then the resistance voltage also doubles and the ratio V_R/V_0 remains constant. This is faster than trying to adjust the output voltage each time to keep it constant.

1. Open the signal generator. Set it to 25 kHz and record this frequency in Table 1.
2. Click on Monitor. If the trace is rolling left or right, click the trigger on the oscilloscope.
3. Adjust the horizontal scale on the scope to show three cycles.
4. Stop monitoring.
5. Adjust the vertical and horizontal scales so that all the peaks are visible.
6. Right click the coordinate tool and set the "Snap to Pixel Distance" to 1. Show 3 significant figures. Enable the delta tool if it is not already enabled.
7. Using the coordinate tool, measure the amplitude of each of the voltages. Record the data in Table 1 (Table 2 if the 3.3kΩ system).
8. Looking at the horizontal axis, find where the two lines have a negative slope, and intersect the V = 0 axis. It may be useful to zoom in to enlarge the signal data.
9. Using the delta tool (coordinate tool + box with cross-hair), measure the difference in time between the two curves. Keep the square box on the VA curve. Put the cross-hair box on the VB curve.
10. The Δx value is the phase shift. Record this value in the Phase Shift Time column in Table 1 (Table 2 if the 3.3kΩ system).
11. Repeat the measurements for each of the frequencies listed on Table 1 (Table 2 if the 3.3kΩ system).
12. Exchange the 1 kΩ resistor for the 3.3 kΩ resistor and repeat the entire procedure. In Table 1 (software), click each of the headers (row 1) and select 3.3 k Ohm instead of 1 k Ohm.
13. Trace both data sets in Figure 3.

Data Collection

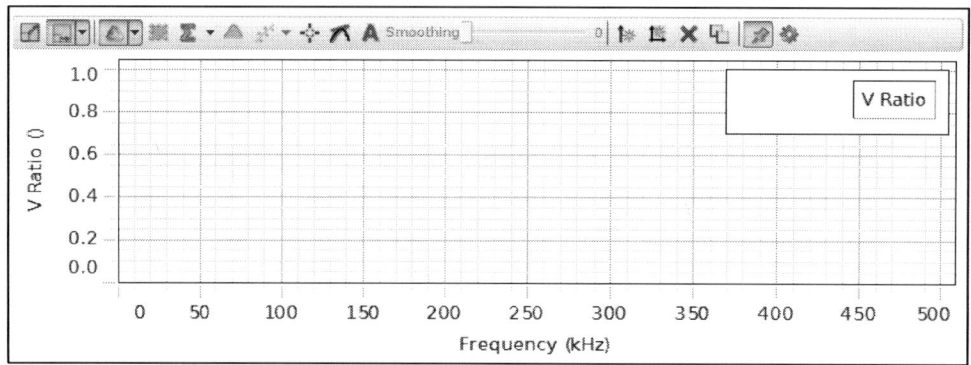

Figure 3: Resonance curve

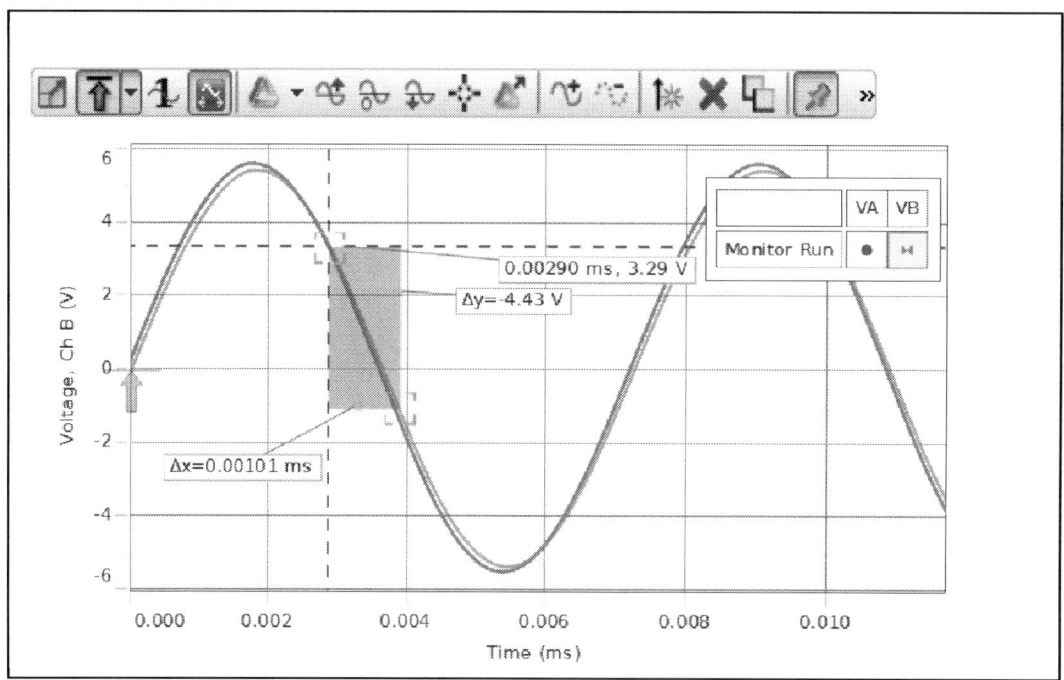

Figure 4: Oscilloscope with Voltages across Resistor and Output

Last/First Name (print):

PHYS 1440-Section   Username ID (xxx9999):

Table 1: Resonance Curve Data for 1kΩ

Frequency (kHz)	Output V_B (V)	Resistor V_A (V)	V Ratio	Phase Shift Time (ms)
25				
75				
100				
125				
130				
135				
140				
150				
175				
225				
300				
400				

Table 2: Resonance Curve Data for 3.3kΩ

Frequency (kHz)	Output V_B (V)	Resistor V_A (V)	V Ratio	Phase Shift Time (ms)
25				
75				
100				
125				
130				
135				
140				
150				
175				
225				
300				
400				

Analysis of the Resonance Curve

1. Display both datasets on the graph in the PASCO capstone file analysis page.
2. Measure the height of the peak for each curve and the frequency of each peak. You might want to expand the horizontal scale. Record them in Table 3.
3. Measure the width of the curve at half the max height for each resistor. As before, you should set the snap to pixel distance to 1 for the coordinate tool.

Table 3: Full Width Half Maximum Analysis

Resistor Value	Resonant Frequency (kHz)		Peak Amplitude	Full Width Half Max at Resonance
1 kΩ				
3.3 kΩ				

Last/First Name (print): _____

PHYS 1440-Section _____ Username ID (xxx9999): _____

Questions:

1. How do the height and width of the Frequency and V Ratio curve change when you increase the resistance?

2. Calculate the theoretical resonant frequencies and compare them to the measured values with a percent difference. Remember that the frequency of the signal generator is f which is related to the theoretical frequency by $f_{res} = \frac{\omega_{res}}{2\pi}$, where $\omega_{res} = \frac{1}{\sqrt{LC}}$.

3. Does the resistance change the resonant frequency?

4. Why isn't the resonant Frequency and V Ratio curve symmetrical about the resonant frequency (why doesn't it look like a standard bell curve)?

1440 Questionnaire: Experiment 11: LRC Circuit-Resonance

Do not put your name on this page. Hand in this page separately.

1. TA name(s):

2. What one thing did you like best about this laboratory?

3. What one thing did you like least about this laboratory?

4. What one thing would you change in this laboratory?

5. What one thing would you leave the same?

Additional Comments?

Last/First Name (print): _____

PHYS 1440-Section _____ Username ID (xxx9999): _____

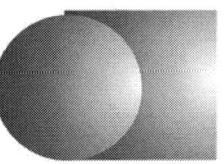

Experiment 12:
Phase Relationships in an LRC Circuit

Pre-lab

1. If an inductor, a capacitor, and a resistor are connected in series with a sine wave generator, what quantity will be common to all three components?

2. How do the capacitor and inductor affect the phase of the voltage?

3. Which component in the LRC circuit has the same phase for both the current and the voltage?

4. At resonance, what is true of the inductive reactance?

Last/First Name (print): _____

Experiment 12:
Phase Relationships in an LRC Circuit

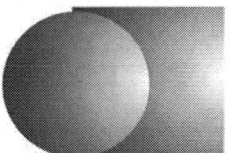

Introduction

The phase differences between the output voltage, the voltage across the inductor, the voltage across the capacitor, and the voltage across the resistor will be examined at resonant frequency. The voltage and phase relationship will also be examined for frequencies above and below resonance.

Theory

An inductor, a capacitor, and a resistor are connected in series with a sine wave generator as shown in figure 1. Since it is a series circuit, the current will be common to all the components and given by

$$I = I_{max}cos(wt)$$ **Equation 1**

The voltage across the resistor will be in phase with the current, but the voltage across the inductor leads the current by 90⁰ and the voltage across the capacitor lags by 90⁰. Adding the three EMF voltages, **E**, results in a total voltage that varies sinusoidally, but has a phase shift φ with respect to the current and is given by

$$\mathbf{E} = \mathbf{E}_{max}cos(\omega t + \varphi)$$ **Equation 2**

The three components obey the AC analogs of Ohm's Law: $E_R = IR$, $E_{L\ max} = I_{max}X_L$, $E_{C\ max} = I_{max}X_C$, where X_L and X_C are the AC analogs of resistance called the inductive reactance and the capacitive reactance. Since the voltage across the resistor is in phase with the current, the phase of the current can be measured by measuring the phase of the voltage across the resistor. The capacitive reactance and the inductive reactance are given by:

$$X_C = \frac{1}{\omega C} \qquad \text{Equation 3}$$

$$\& \quad X_L = \omega L \qquad \text{Equation 4}$$

In terms of the measured variables:

$$\frac{V_C}{V_R} = \frac{1}{\omega C R} \qquad \text{Equation 5}$$

$$\& \quad \frac{V_L}{V_R} = \frac{\omega L}{R} \qquad \text{Equation 6}$$

At resonance, $X_L = X_C$ and $\varphi=0$. Setting equation 3 equal to equation 4 yields the resonant angular frequency:

$$\omega_{res} = \frac{1}{\sqrt{LC}} \qquad \text{Equation 7}$$

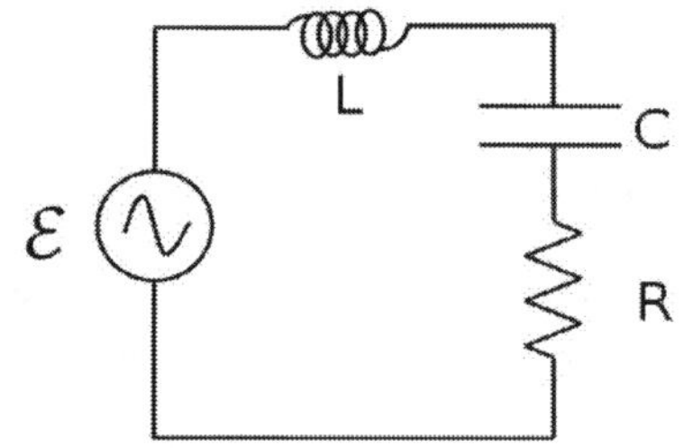

Figure 1: LRC Series Circuit

Phasors:

A convenient way to summarize the information on the Theory page is to use phasors. Recall that circuit current is in phase with V_{Ro}. At resonance, $V_C = V_L$ and they cancel each other. So the total voltage (ε_{max}) equals the voltage across the resistor (V_{Ro}) and the phase shift is zero. Below resonance (the case shown), $V_C > V_L$, the capacitor dominates, total voltage lags total current and the phase shift is negative. Above resonance the inductor dominates.

The magnitudes of the phasors are related through phasor math where ε_{max} is the sum of the other three as indicated in Figure 2 where the phasors are added like vectors.

Measuring the Phase Angle:

We will measure the time Δt by which a signal leads or lags the circuit current (actually V_R). Since 360° of phase shift is equivalent to one cycle, we have:

$$\varphi = 360° \left(\frac{\Delta t}{T}\right) = 360° (\Delta t) f \qquad \textbf{Equation 8}$$

where T is the period and f is the frequency.

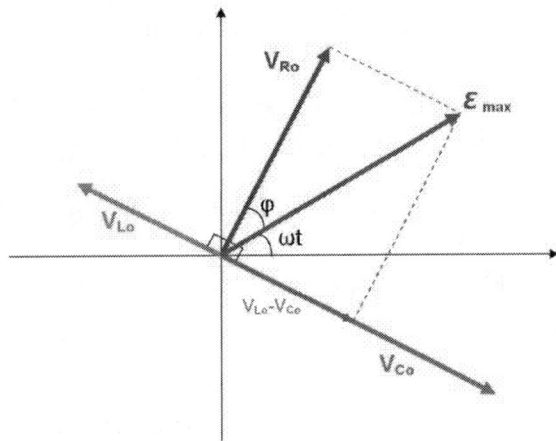

Figure 2: Phasors

Equipment

1	Resistor/Capacitor/Inductor Network	UI-5210
4	Voltage Sensors	UI-5100
1	BNC-to-Banana Cord for 850 Output	UI-5119
1	Patch Cords (Set of 5)	SE-7123
1	PASCO Capstone	UI-5400
1	850 Universal Interface	UI-5000

Setup

1. Construct the circuit shown in figure 3 using the following values. Figure 4 shows the constructed circuit.
   $$L = 6.8 \text{ mH}$$
   $$C = 3900 \text{ pF}$$
   $$R = 1.0 \text{ k}\Omega$$

2. Connect to Output #2 on the 850 Universal Interfaces using the BNC to banana cord. Note that it is important to connect the left hand side of the inductor to the right hand side of the capacitor and the left hand side of the capacitor to the right hand side of the resistor as shown. If the cords are connected in reverse it will cause the voltages to be inverted.

3. Connect a voltage sensor to the D analog input on the 850 Universal Interface. Connect the leads over the input so that it measures the total voltage ($V_D=V_O$) as shown in Figure 4.

4. Connect the other three voltage sensors as shown in Figure 5. Input A is connected across the resistor. Input B is connected across the inductor. Input C is connected across the capacitor. The black side of each voltage sensor is attached on the left side of the circuit board.

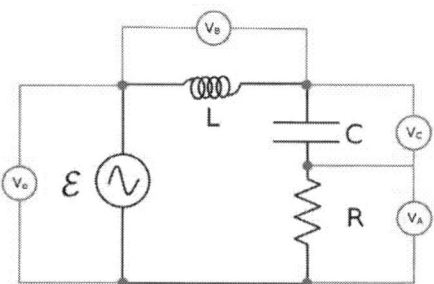

Figure 3: Circuit Diagram with Voltage Sensors

Figure 4: LRC Circuit

Figure 5: Circuit with Sensors

Last/First Name (print): _____

PHYS 1440-Section _____ Username ID (xxx9999): _____

Procedure

1. Open the Signal Generator.
2. Set signal Generator #2 to a frequency of 30000 Hz and an amplitude of 6 V. Select Auto.
3. Click Monitor. The scope trace obtained is shown in figure 6. If the scope trace is rolling left or right, click the trigger button in the scope toolbar.
4. Adjust the vertical scale so the signals are as large as possible without going off screen.
5. Adjust the horizontal scale so you see two cycles.
6. Click "Automatically adjust sample rate based on time axis" button. It is a green button on the graphs tool bar, with a yellow sine wave and two arrows connected by a red line.
7. Adjust the Signal Generator Frequency (to the nearest 100 Hz) to make the phase $\varphi=0$ by making V_D (=V_O) and V_A (=V_R) cross the V=0 axis at the same point.
8. Adjust the Signal Generator Amplitude so that the peak voltage of V_R exactly 6 V.
9. Click Stop. .Record the frequency (resonance) in Table 1 on the Conclusion page.
10. Export and Rename each dataset in the data summary tab as "resonance"
11. Click Monitor. Open Signal Generator and adjust the Signal Generator Frequency below resonance so the peak of $V_R = 4$ V.
12. Adjust the horizontal scale for two cycles.
13. Click Stop.
14. Repeat step 4 above; this time re-label as "below res."
15. Record the frequency in Table 1.
16. Repeat for the $V_R = 4$ V above resonance. The signal will be jagged since at 1 MHz there are only about 20 points per cycle.

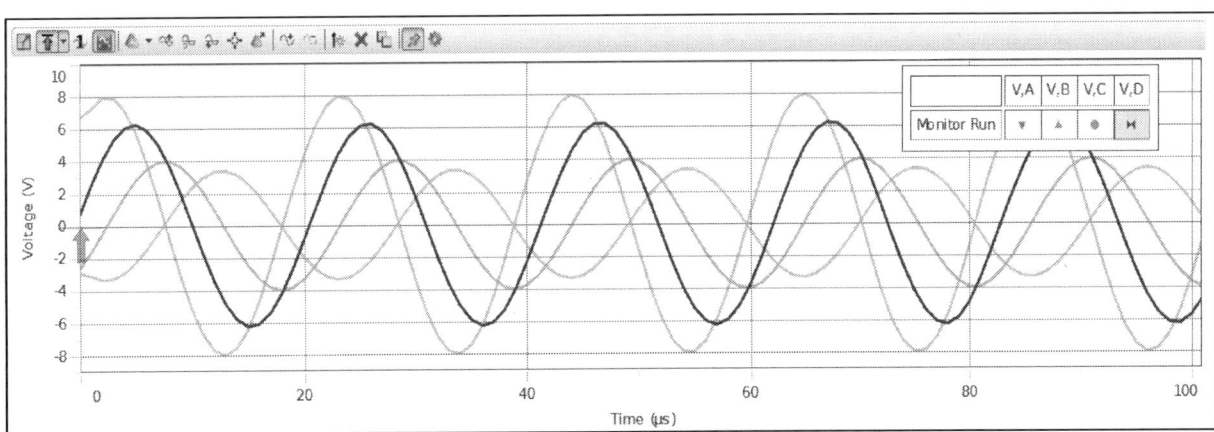

Figure 6: Scope output showing voltage vs. time

Analysis: Phase Relationships

1. Select the four "resonance" data sets as shown in figure 7. Expand the horizontal scale so 1 cycle shows. Set the vertical scale so the range is from -0.5 V to 2.5 V. This makes the V=0 crossings easy to see.
2. Right click on the Coordinate tool and set the "snap distance" to 1. Change the significant figures to 3.
3. To measure the phase shift of the voltage, the delta tool will be used. The delta tool has two boxes. The **box without a cross-hair** will be positioned on the first place V_R (V_A) crosses the time axis with a negative slope. To measure Δt VL, take the **box with a crosshair** and measure to the nearest VL (V_B) intersection with a negative slope. Record the Δx value in the Δt VL cell. Repeat this process for the other voltages, keeping the reference point (box without a cross-hair) on the V_R (VA) intersection and measuring with the cross-hair box. The cross-hair box may go before or after (within 360°) the reference point (V_R, VA).
4. The phase shift can be calculated using equation 8, or by plugging in the values in the computer software
5. Repeat for the "below res" and "above res" data.

Note: ($V_A = V_R$, $V_B = V_L$, $V_C = V_C$, $V_D = V_O$)

Table 1: Phase Data

System	Frequency (Hz)	Δt VL (μs)	Δt VC (μs)	Δt Vo (μs)	φL (deg)	φC (deg)	φtot (deg)
Resonance							
Below res							
Above res							

Last/First Name (print): _____

PHYS 1440-Section _____ Username ID (xxx9999): _____

Analysis: Voltage

1. Select the four "resonance" data sets
2. Right click on the Coordinate tool and select Tool Properties. Change "Snap to Pixel Distance" to 1.
3. Use the Coordinate tool to determine the maximum voltages: V_R, V_L, V_C, V_0 and enter them in the Table 2.
4. Repeat for the "below res" and "above res" datasets.
5. Calculate the remaining 4 columns. VL/VR and VC/VR will be calculated using the experimental values found. Calculate the theory values using equations 5 and 6.
6. For the "above resonance" data that is still in the analysis screen, pick any vertical line of time, and record the voltage at each point on that line. Record each voltage in Table 3.

Table 2: Voltage Summary

System	Frequency (Hz)	VR (V)	VL (V)	VC (V)	Vo (V)	VL/VR	VC/VR	VL/VR Theory	VC/VR Theory
Resonance									
Below Res									
Above Res									

Table 3: Test of Kirchhoff's Voltage Law for a Unit of Time

V_R = VA	V_L = VB	V_C = VC	V_0 = VD

Last/First Name (print): _____

PHYS 1440-Section _____ Username ID (xxx9999): _____

Conclusions:

1. Using equation 7, calculate the resonant angular frequency and the resonant frequency and compare to experimental value for resonance.

2. Does $X_C = X_L$ at resonance? What are the values of X_C and X_L at the particular resonance found for this circuit? Equations 3 and 4 will help.

3. Which reactance (X_C or X_L) is larger above resonance? Which reactance is larger below resonance?

4. Considering the total phase shift, "φ tot", does the circuit behave more like a capacitor or more like an inductor below resonance? The last column of Table 1 will help.

5. Using the data from Table 2, do the voltages at a given frequency obey scalar math? That is, do the peaks of $V_R+V_C+V_L=V_o$? Look at "below resonance", "resonance", and above "resonance".

6. Does the data in Table 3 imply that for an instant in time, Kirchhoff's voltage law is always true?

1440 Questionnaire: Experiment 12: Phase Relationships in an LRC Circuit

Do not put your name on this page. Hand in this page separately.

1. TA name(s):

2. What one thing did you like best about this laboratory?

3. What one thing did you like least about this laboratory?

4. What one thing would you change in this laboratory?

5. What one thing would you leave the same?

Additional Comments?

Last/First Name (print): _____

PHYS 1440-Section _____ Username ID (xxx9999): _____

Experiment 13:

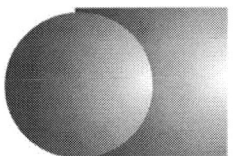

Reflection, Refraction, Dispersion of Light, and Brewster's Angle

Pre-lab, Part A

1. If a ray of light strikes a plane mirror at an angle that is 33° from the line normal to the surface of the mirror, what is the angle of reflection Θ_r?

Pre-lab, Part B

2. As light travels from a low to a high index of refraction, what happens to its speed?

3. If a ray of light initially in air (n = 1.00), strikes a block of ice (n = 1.31) at an angle that is 33° from the normal, what is the angle of refraction Θ_2?

219

Pre-lab, Part C

4. Define dispersion of light.

Pre-lab, Part D

5. What condition regarding the relationship between angle of reflection, Θ_r, and angle of refraction, Θ_R, allows for Brewster's angle to occur? Give the equation to calculate this angle as well.

Last/First Name (print): _____

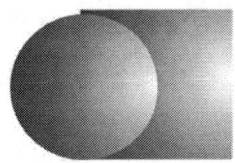

Experiment 13:

Reflection, Refraction, Dispersion of Light, and Brewster's Angle

Introduction

This experiment will examine the 4 optical phenomenon mentioned in the experiment title. The student will obtain a more intuitive understanding of light in the context of reflection, refraction, dispersion of light, and Brewster's angle.

Theory A: Reflection

When a ray of light (incident ray) strikes a plane mirror, the light ray reflects (reflected ray) off the mirror and changes its direction of travel. The angle of incidence θ_i is the angle between a line normal to the surface and the incident ray, as in Figure A1; the angle of reflection θ_r is the angle between the normal and the reflected ray. The law of reflection states that the angle of incidence equals the angle of reflection:

$$\theta_i = \theta_r \qquad \textbf{Equation A1}$$

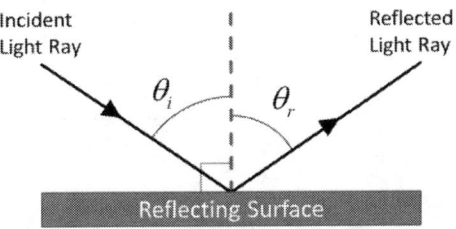

Figure A1: Reflecting Light

221

Equipment

1	Ray Optics Kit	OS-8516
1	Basic Optics Light Source	OS-8470
1	Protractor	
1	Ruler	
1	White paper	
1	Pencil	
1	D-Shaped Acrylic Lens	w/OS-8465
1	Polarizer	OS-8533A

Setup A: Reflection

1. Place a sheet of white paper on the lab bench.
2. Set the three-in-one mirror in the center of the paper.
3. Use the pen or pencil to trace the outline of the mirror on the paper.
4. Lift the mirror off the paper.
5. Use the ruler to measure half the length of the flat side of the outline.
6. Make a small mark at the half-way point.
7. Use the ruler to draw a straight line from the apex opposite the flat side of the outline, through the half-way mark you just made.
8. Extend the line several centimeters past the outline as in Figure A2. This line represents a line normal to the flat mirror surface.
9. Use the ruler to measure 1.8 cm inside the outline from the flat side of the outline following the line you just drew.
10. Make a small mark as in Figure A3.
11. Use the ruler to draw a straight line from the apex opposite the concave side of the outline, through the center mark you just made and then through the concave side of the outline.
12. Extend the line several centimeters past the outline as in Figure A4. This line represents a line normal to the concave mirror surface.
13. Use the ruler to draw a straight line from the apex opposite the convex side of the outline, through the center mark and then through the convex side of the outline.
14. Extend the line several centimeters past the outline as in Figure A4. This line represents a line normal to the convex mirror surface

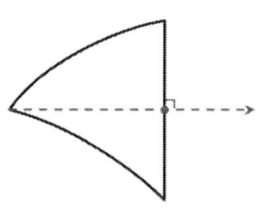

Figure A2

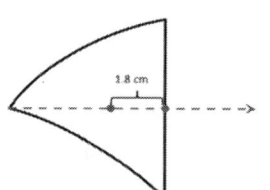

Figure A3

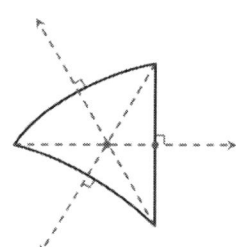

Figure A4

Last/First Name (print): _____

PHYS 1440-Section _____ Username ID (xxx9999): _____

Procedure A1: Flat Mirror Reflection

1. Place the three-way mirror on the paper and align it inside the outline.
2. Place the basic optics light source flat on the lab table.
3. Rotate the knob on the front of the light source to produce a single ray of light as in Figure A5.
4. Adjust the position of the light source so the single light ray hits the flat mirror at the point the normal line extends from the outline as in Figure A5.
5. Use the ruler and pen or pencil to trace the paths of the incident and reflected light rays.
6. Label each line "Ray 1 Incident" or "Ray 1 Reflected" similar to Figure A5 to help differentiate them.
7. Lift the mirror off the paper and then use the protractor to measure the angles of both incident and reflected rays relative to the normal line.
8. Record both values into Table A1.
9. Repeat the same procedure two more times using two different incident angles.
10. Record all data into Table A1

Table A1

Flat Mirror	
Angle of Incidence (degrees)	Angle of Reflection (degrees)

Figure A5: Flat Mirror Reflection

Procedure A2: Concave Mirror Reflection

1. Adjust the position of the light source so the single light ray now hits the concave mirror at the point the normal line extends from the outline as in Figure A5.
2. Use the ruler and pen or pencil to trace the paths of the incident and reflected light rays.
3. Label each line "Ray 1 Incident" or "Ray 1 Reflected" similar to Figure A5 to help differentiate them.
4. Lift the mirror off the paper.
5. Use the protractor to measure the angles of both incident and reflected rays relative to the normal line.
6. Record both values into Table A2.
7. Repeat the same procedure two more times using two different incident angles.
8. Record all data into Table A2.

Table A2

Concave Mirror	
Angle of Incidence (degrees)	Angle of Reflection (degrees)

Procedure A3: Convex Mirror Reflection

1. Adjust the position of the light source so the single light ray now hits the convex mirror at the point the normal line extends from the outline as in Figure A5.
2. Use the ruler and pen or pencil to trace the paths of the incident and reflected light rays.
3. Label each line "Ray 1 Incident" or "Ray 1 Reflected" similar to Figure A5 to help differentiate them.
4. Lift the mirror off the paper
5. Use the protractor to measure the angles of both incident and reflected rays relative to the normal line.
6. Record both values into Table A3.
7. Repeat the same procedure two more times using two different incident angles.
8. Record all data into Table A3.

Table A3

Convex Mirror	
Angle of Incidence (degrees)	Angle of Reflection (degrees)

Last/First Name (print): _____

PHYS 1440-Section _____ Username ID (xxx9999): _____

Analysis A: Reflection

1. What is the relationship between the angle of incidence and the angle of reflection?

2. Does it matter whether the reflective surface is flat or curved? Why?

3. What are sources of experimental uncertainty for this part of the experiment?

Theory B: Refraction

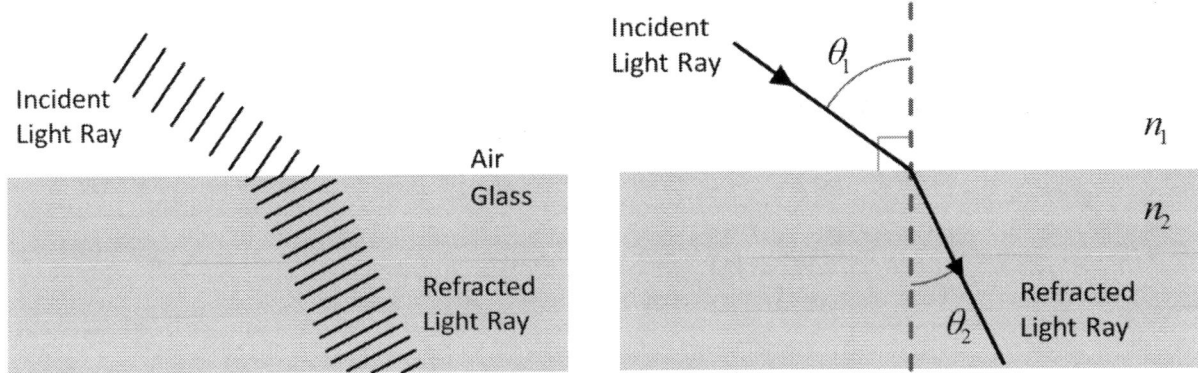

Figure B1: Refraction

Refraction is the bending of the path of light when it passes from one material to another. The most common explanation of refraction includes the wave theory of light: Imagine a ray of light traveling through air. When this ray strikes the surface of a piece of glass at an angle, one side of the wave front enters the glass before the other and it slows down, while the other side continues to move at its original speed until it too reaches the glass. As a result, the light ray bends inside the glass. The amount at which a medium (like glass) can slow a light ray down is based on the medium's index of refraction n. If a medium has a high index of refraction (like glass, $n \sim 1.4$), light will travel slower through it. If a medium has an index of refraction close to 1 (like air, $n \sim 1$) light will travel faster through it.

In addition to the different indexes of refraction, the angle at which a light ray bends when traveling from one medium to another is also affected by the angle at which the original light ray is incident. The relationship is described by a simple equation known as Snell's Law (Eq.2). Where n_1 is the index of refraction of the first (incident) medium, n_2 is the index of refraction of the second medium, and θ_1 and θ_2 are the incident and refracted angles respectively. This is known as Snell's Law.

$$n_1 \sin \theta_1 = n_2 \sin \theta_2 \qquad \textbf{Equation B1}$$

Rearranging this equation to solve for n_2 yields the following:

$$n_2 = n_1 \frac{\sin \theta_1}{\sin \theta_2} \qquad \textbf{Equation B2}$$

Setup B: Refraction

1. Place the basic optics light source flat on the lab table.
2. Rotate the knob on the front of the light source to produce a single ray of light as in Figure B2.
3. Place the basic optics ray table on the lab bench just in front of the light source, such that the light ray passes through the center of the ray table.
4. Rotate the top of the ray optics table so that light ray is aligned with the 0° mark on the tabletop.
5. Place the D-shaped acrylic lens in the center of the table (frosted side down).
6. Adjust its position so it fits within the lens outline on the tabletop.

Figure B2: Refraction Setup

Figure B3: Measuring Refraction Angle

Last/First Name (print): _____

PHYS 1440-Section _____ Username ID (xxx9999): _____

Procedure B: Refraction

1. Set the incident angle to 0° by rotating the top of the basic optics ray table. Make certain that the light ray hits the flat surface of the lens at its center.
2. Observe the refracted ray of light and record the refraction angle in Table B1.
3. Repeat the same procedure for each incident angle in Table B1.
4. Record all measurements into Table B1 and make certain that the light ray hits the flat surface of the lens at its center for all measurements.

Table B1: Refraction Angles for D-Shaped Lens

Incidence Angle (degrees)	Refraction Angle (degrees)
0	
10	
20	
30	
40	
50	
60	
70	

Analysis B: Refraction

Plot a graph showing $\sin \theta_1$ versus $\sin \theta_2$ for your data. Equation 2 will be used to solve for n_2.

1. Use a linear fit to plot the data and find slope of the line to help determine a value for the index of refraction for the acrylic lens (n_2). Assume that the index of refraction for air (n_1) is 1. Record your experimental value here:

$n_2 =$

2. If acrylic glass has a theoretical index of refraction of 1.49, what is the percent error on your experimental value?

$$\%\text{error} = \frac{|\text{Theoretical} - \text{Experimental}|}{\text{Theoretical}} \times 100\%$$

%error =

3. What caused error in these measurements and how can they be minimized?

Theory C: Dispersion

Ordinary white light is a combination of waves with wavelengths extending over the visible spectrum (from red to violet). The speed of light in a vacuum is the same for all colors of light, but the speed in a material is different for different wavelengths. Therefore, the index of refraction of a material depends on the wavelength (color) of the light that passes through the material, and is a property of the incident light as well as a property of the material. If a beam of light contains more than one color of light, each color will refract by a different amount and each color will come out of the material traveling in a different direction. This is called dispersion: the separation of a beam of light into its component colors by refraction. When ordinary white light passes through a material, the dispersion is observed as a rainbow that comes out the other end.

The velocity of light is related to the speed of light and the index of refraction

$$v = \frac{c}{n} \qquad \text{Equation C1}$$

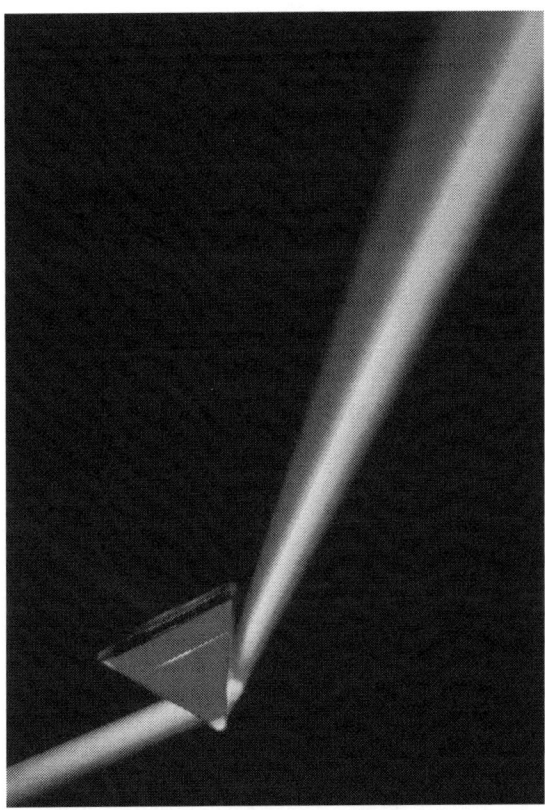

Figure C1: Dispersion from a prism

Setup C: Dispersion

1. Place a sheet of white paper on the lab bench.
2. Set the acrylic rhomboid (frosted side down) near the top-left corner of the paper.
3. Adjust the orientation of the rhomboid to match that in Figure C2.
4. Use the pen or pencil to trace the outline of the rhomboid on the paper.
5. Lift the rhomboid off the paper.
6. Choose a point (any point) along the angled side of the outline.
7. Use the protractor to draw a normal line at that point similar to Figure C3.
8. At the same point, draw a 45°-line relative to normal that extends outward from the outline similar to Figure C3.
9. Place the rhomboid back on the paper (frosted side down).
10. Adjust its position so that it sits within the outline.
11. Place the basic optics light source flat on the lab table.
12. Rotate the knob on the front of the light source to produce a single ray of light as in Figure C4.
13. Adjust the position of the light source so that the light ray is incident on the angled side of the rhomboid.
14. Aligned with the 45° line drawn on your paper.

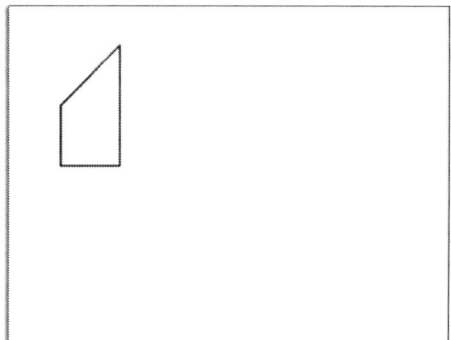
Figure C2: Rhomboid on Paper

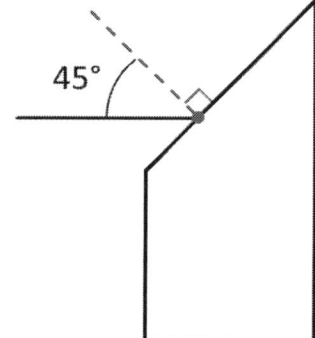
Figure C3: Rhomboid with Lines

Figure C4: Single Light Ray

Procedure C: Dispersion

1. Observe the refracted light ray as it passes through the rhomboid and travels down the length of the paper.
2. Use the pen or pencil to carefully mark on the outline where the light ray exits the dispersion interface. The surface of the rhomboid at which the light ray leaves is called the "dispersion interface."
3. Follow the dispersed ray of light as far down the paper as possible without going off.
4. Carefully make a small mark where the blue light and red light reach the end of the paper Similar to Figure C5.
5. Remove the rhomboid from the paper.
6. Use the ruler to draw a line between the point at which the incident ray hit the rhomboid and where it exited at the dispersion interface.
7. Use the ruler to draw a line connecting the point at which the light ray exited the dispersion interface and each color mark (red and blue) at the end of the paper.
8. Label each line "red" or "blue" to differentiate them.
9. Finally, use the protractor to draw a line normal to the dispersion interface at the point from which the light ray exited, that bisects the dispersion interface, as in Figure C6.

Figure C5: Marking the Paper

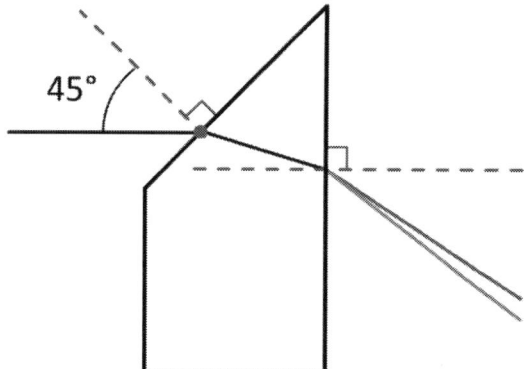

Figure C6: Dispersion Lines

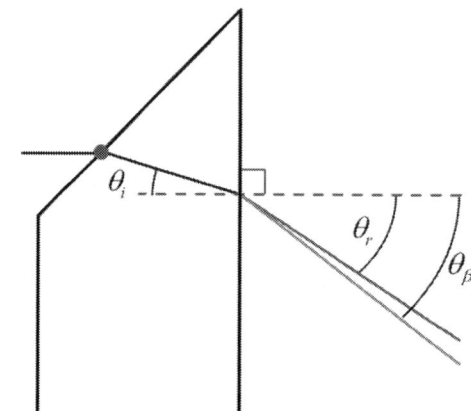

Figure C7: Dispersion Angles

Important Note: Figures C6 and C7 are simplified. At each interface, or boundary, dispersion will occur. The light begins to disperse when it enters the prism and continues to disperse after is exits the prism. Because of how small the initial dispersion is and the limitation of the precision of the experiment we are able to ignore the dispersion from within the prism.

Last/First Name (print): _____

PHYS 1440-Section _____ Username ID (xxx9999): _____

Analysis C: Dispersion

1. Use the protractor to measure the incident angle at the dispersion interface θ_i, as well as the dispersion angles θ_r for red light and θ_β for blue light as seen in Figure C7. Record measurements below for all angles in terms of degrees (°).

    $\theta_i = $ _____

    $\theta_r = $ _____

    $\theta_\beta = $ _____

2. Based on the two different dispersion angles of red light and blue light, which color traveled faster in the acrylic rhomboid? How do you know?

3. Use Snell's Law to calculate the index of refraction for the two different colors of light (n_r for red and n_β for blue) in the acrylic rhomboid. Assume that the light ray basically remains a single white light ray (does not disperse much) as it travels through the rhomboid, and the index of refraction for air is 1.

    $n_r = $ _____

    $n_\beta = $ _____

4. Given the definition of index of refraction, and your values from the previous question, calculate the speed of light for each color (v_r for red and v_β for blue) in the acrylic rhomboid. Record the results below. Equation 1 will help. This velocity cannot be faster than the speed of light in a vacuum ($\approx 3 \times 10^8$ m/s)!

$v_r = $ _____

$v_\beta = $ _____

5. Describe the relationship between wavelength of light and the speed at which it travels through a medium (like acrylic).

D: Brewster's Angle

When unpolarized light is incident on a transparent dielectric surface, most of the light is refracted through the transparent medium, but some light is also reflected, as in Figure D1. The amount of light reflected is dependent on the angle of incidence and the polarization of the incoming light.

In this experiment changes in intensity and polarization of light will be explored as the light changes when reflected from the surface of a transparent medium. The results will be compared to a theoretical value known as Brewster's Angle.

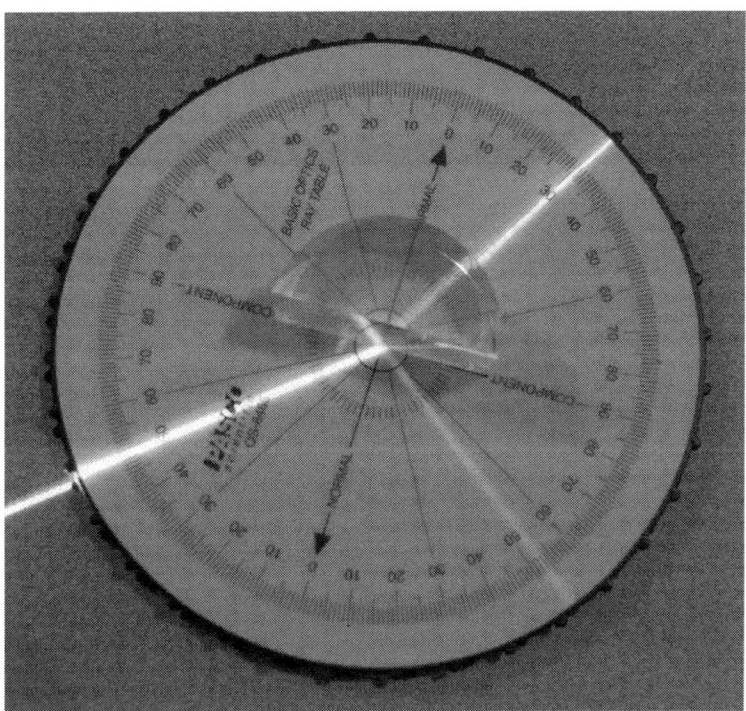

Figure D1: Light being Refracted and Reflected

Theory D: Brewster's Angle

As you have observed, when unpolarized light reflects off a nonconducting surface like the acrylic D-shaped lens, it is partially polarized parallel to the plane of the reflective surface depending on the incident angle of the incoming light. As you have also seen, there is a specific incident angle at which the reflected light is 100% polarized. This angle is called Brewster's angle, and it occurs when the reflected ray and the refracted ray are 90° apart.

To determine the mathematical relationship that describes Brewster's angle we can start by using Snell's Law:

$$n_1 \sin \theta_P = n_2 \sin \theta_2 \qquad \textbf{Equation D1}$$

Because, $\theta_P + \theta_2 = 90°$ or $\theta_2 = 90° - \theta_P$, an important substitution can be made into Eqn.1:

$$n_1 \sin \theta_P = n_2 \sin(90° - \theta_P) \qquad \textbf{Equation D2}$$

Substituting $n_2 \sin(90° - \theta_P) = \cos \theta_P$ back into Eq.2 gives:

$$n_1 \sin(\theta_P) = n_2 \cos \theta_P \quad \text{or} \quad \tan \theta_P = \frac{n_2}{n_1} \qquad \textbf{Equation D3}$$

Equation 3 describes the relationship between indices of refraction, n_1 and n_2, and Brewster's angle θ_P.

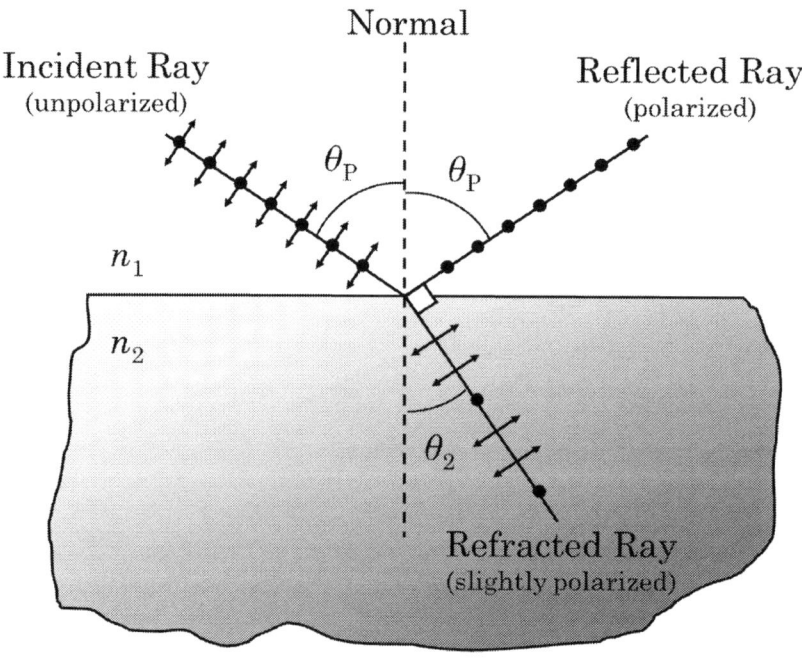

Figure D4: Brewster's Angle

Setup D: Brewster's Angle

1. Set the Ray table on the lab bench.
2. Place the clear acrylic D-shaped lens within the outline on the top of the ray table.
3. Connect the light source to power
4. Adjust the dial on the front of it so that only one ray of light is emitted.
5. Turn off all the lights in the classroom.
6. Place the light source on the lab bench so that the front of it is as close to the ray table as possible without touching it, and the ray of light is incident on the center of the flat side of the D-shaped lens (similar to Figure D2).
7. Adjust the incident angle of the light ray to 25° making sure to keep the light ray incident on the center of the flat side of the D-shaped lens.
8. Make a note below about what percentage of the incident light appears to be reflected assuming all of the light that is incident on the D-shaped lens is either refracted or reflected.

9. Respond to the following in the space below. Is the incident light ray polarized? If not, what could you do to make it polarized?

10. Increase the incident angle to several larger angles.
11. Make a note below about what happens to the apparent percentage of incident light that is reflected as the incident angle increases.

12. Adjust the incident angle back to 25°.

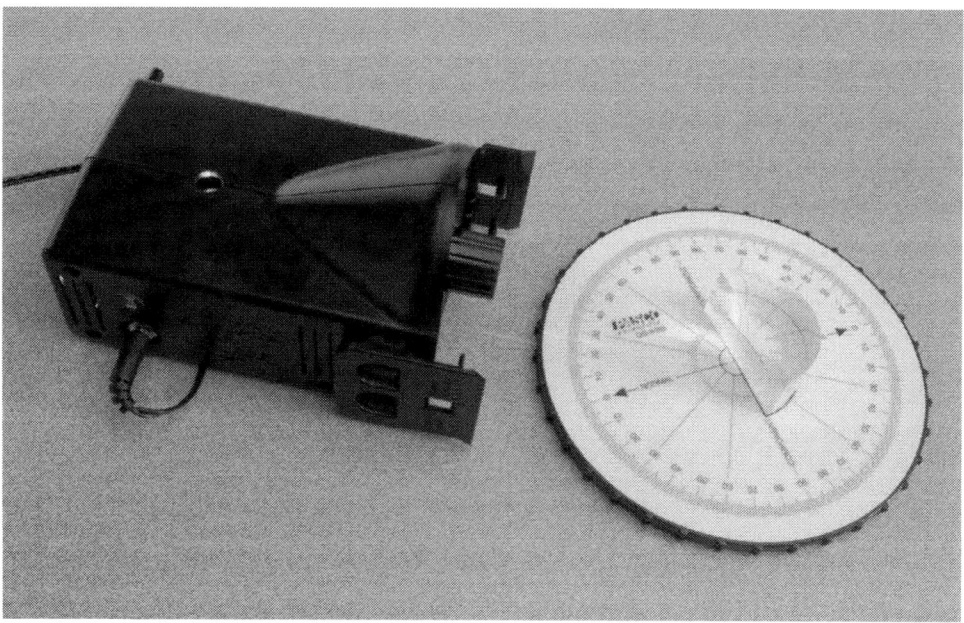
Figure D2: Ray Table Setup

Data: Brewster's Angle

1. Close one eye and look down the path of the reflected light ray (as in Figure D3) observing the intensity of the reflected ray on the surface of the ray table.
   NOTE: The plane of the lab table should be at eye level. It may be useful adjust the position of your head side-to-side to best see the reflected light ray.

2. Place the polarizer disk in front of your open eye and slowly rotate the disk one full revolution. Make a note below about what happens to the intensity of the reflected ray observed on the surface of the ray table as you rotate the polarizer disk.

3. Adjust the incident angle to 40° and repeat the previous steps. Make a note below about what happens to the intensity of the reflected ray observed on the surface of the ray table as you rotate the polarizer disk?

4. What happened to the polarization of the reflected light ray when the incident angle was increased? How does the polarizer disk indicate the polarization of the reflected light ray?

5. Observe the reflected light ray using the polarizer disk for each incident angle between 45° and 75°, using 5° increments. Be sure to rotate the polarizer disk one full revolution for each incident angle and pay close attention to the intensity of the reflected light ray.

6. Describe your observations of the reflected light ray at increasing incident angles. What changed as incident angle increased, and what stayed the same?

7. Use the polarizer disk and ray table to determine the incident angle and polarizer orientation (rotation) that allows none of the reflected light ray to be seen. Record the incident angle below.

Figure D3: Viewing the Reflected Light

Last/First Name (print): _____

PHYS 1440-Section _____ Username ID (xxx9999): _____

Analysis D: Brewster's Angle

1. Calculate the theoretical value for Brewster's angle using the D-shaped acrylic lens. Assume that the index of refraction of air is 1 and index of refraction for the acrylic lens is 1.49.

2. How does your experimental value for Brewster's angle from the Collect Data page compare to the theoretical? What is the percent error?

$$\%\text{error} = \frac{|\text{Theoretical} - \text{Experimental}|}{\text{Theoretical}} \times 100\%$$

1440 Questionnaire: Experiment 13: Reflection, Retraction, Dispersion and Brewster's Angle

Do not put your name on this page. Hand in this page separately.

1. TA name(s):

2. What one thing did you like best about this laboratory?

3. What one thing did you like least about this laboratory?

4. What one thing would you change in this laboratory?

5. What one thing would you leave the same?

Additional Comments?